Governance and Management for Sustainable Water Systems

Governance and Management for Sustainable Water Systems Series

Governance and Management for Sustainable Water Systems

Neil S Grigg

IWA Publishing
London • New York

Published by **IWA Publishing**
Alliance House
12 Caxton Street
London SW1H 0QS, UK
Telephone: +44 (0)20 7654 5500
Fax: +44 (0)20 7654 5555
Email: publications@iwap.co.uk
Web: www.iwapublishing.com

First published 2011

Cover design by www.designforpublishing.co.uk
Typeset by MPS Limited, a Macmillan Company, Chennai, India, www.macmillansolutions.com
Index provided by Georgina Bowden

British Library Cataloguing in Publication Data
A CIP catalogue record for this book is available from the British Library

Library of Congress Cataloging-in-Publication Data
A catalog record for this book is available from the Library of Congress

ISBN13: 9781843393467
ISBN: 1843393468

Contents

Acknowledgments

In approaching the subject of water governance, I have had the good fortune to work with incredible people along the way. The first group comprises the dedicated people in the international community. They are too many to name, but they work in the UN system, the World Bank, U.S. agencies, donor organizations, and other groups who are dedicated to improving the world. Sometimes their understanding of issues amazes me.

Opportunities come at unexpected times, and I want to thank Walton Jones and David Howells, who arranged for me to direct the University of North Carolina Water Research Institute and to serve in important roles of state government. More than any other experience, this one helped me with insight into water governance.

Early in my academic career, Murray McPherson and Scott Tucker, the Director and Deputy Director of the Urban Water Resources Research Program of the American Society of Civil Engineers taught me about integration in urban water systems. Scott showed me through his long tenure as Executive Director of Denver's Urban Drainage and Flood Control District about effective management and inter-jurisdictional coordination in a regional setting.

Dan Okun of the University of North Carolina was an amazing and dedicated professional. His first lesson was about the regionalization of water management in the UK, about which he published a book, and he taught me later about water and sanitation in developing countries and about important aspects of water distribution systems.

My colleagues at Colorado State University have inspired me with many lessons about water management in developing countries. First and foremost, Maurice Albertson was a long term inspiration to focus on development and poverty. My colleague Evan Vlachos has provided insight into many institutional issues, such as institutional mobilization, and he has served as a great conduit for information about integrated water management around the world.

Many of our graduate students have contributed to my understanding. In particular, I want to thank Yoshihiro Katsuhama for sharing experience on international river basins and capacity-building for flood control.

I am so thankful to my educational institutions for awakening me to international issues. These include the US Military Academy at West Point and Colorado State University, in particular. USMA teaches foreign languages and opens avenues for travel, while Colorado State is a fine example of a land grant university with global outreach.

Although I visit developing countries, I have never actually lived in one of them and I respect their experiences and cultural issues. I witnessed the rural South of the U.S. during the 1940s and 1950s, where living conditions were similar and I salute those who escaped those conditions and went on to better lives. I want to mention in particular George Rowe, who was public works director of Cincinnati and became the second African-American President of the American Public Works Association. I served with him on its board of directors in 1994–95 and thought he exemplified the kind of public administrator we need.

Finally, I wish to thank Maggie Smith, Editor of IWA Publishing, who has made my work in publishing this book both easier and more pleasant.

Neil S Grigg

Foreword

While our global society faces many tough issues, water seems to rate high on almost every list. No wonder. With the ever-growing world population set to move past 7 billion soon, it's difficult to provide even basic services in many places. And natural systems are under pressure from drought, pollution, climate change, or other stressors.

The Economist in a special report on May 22, 2010 explained the problem with this phrase about water: "Finite, vital, much wanted, little understood, water looks unmanageable" (Grimond, 2010).

Water problems won't be solved by marketplace allocation or government decisions alone. Mostly, they require the application of good governance. However, as the World Water Assessment Programme (2007) wrote: "In many countries, water governance is in a state of confusion: in some countries there is a total lack of water institutions, and others display fragmented institutional structures or conflicting decision-making structures."

Water governance is no longer limited to government. Today, there is broad acceptance of the involvement of private industry and civil society institutions.

There is much interest in water governance because many experts see it as the best answer we have for today's challenges. In launching this book series on Governance and Management for Sustainable Water Systems IWA Publishing saw the need for reliable guidance to explain the elements of effective water governance systems.

To prepare this introductory volume, I found a rich lode of knowledge to draw from, particularly from international institutions. The literature on general concepts such as integrated water resources management is abundant, and many texts and papers explain the fine points of water law and policy. I was also able to draw from my own 40 years of experience in different water management roles and from the many case studies prepared by others.

The main premise of the book is that water stewardship requires an effective combination of management and governance. The governance framework outlined in the book supports a comprehensive management model for balanced water uses based on the triple bottom line of economic, social, and environmental values. The book addresses the simultaneous challenges of managing water, handling conflicts, and serving environmental needs.

The chapters range across many subjects, but they recognize that governance is not a static concept to be applied in the same way everywhere. It must conform to the context in which it is to be applied. After all, governance is not just about regulatory controls, its job is to see what society's needs are met.

The book could not cover all of the elements of governance in detail and some of them, like regulation for safe drinking water, are described detail elsewhere. Other topics, such as balancing instream flows and trading off agriculture and ecosystem services, need more exposure.

To strip away unnecessary clutter, the book seeks to avoid overly-complex explanations and drill directly to policy, control, and empowerment as elements of water governance.

The main audience for the book comprises the people who have or will have important roles in water governance. These include: water managers who can see the big picture of water integration; boards of directors who must exercise corporate governance over their organizations; policy makers who aim to reform water laws; regulators who need guidance for their decisions; donors who must set investment policies; and students who will lead as future water managers.

My own experience has included a number of these roles and I've tried to write for others in similar situations. As a river master, I currently administer water deliveries under an interstate water compact and U.S. Supreme Court decree. This role is like a water judge with limited powers in assessment, adjudication, and enforcement. In the past, I was responsible for a state government department that issued and enforced environmental permits. I have been a consultant in many legal and political situations related to water, and I've served on several local water boards. With these experiences, I have developed several graduate courses about water management and have advised graduate students working on water issues across the U.S. and in other countries. The graduate students have prepared theses about state water use systems, interstate compacts, regulation, and reform of water governance in developing countries.

I've tried to weave the background knowledge, case studies, and experiences into a coherent explanation of water governance, but it is a complex subject and requires much more analysis than it is given here. Above all, it involves application of the best principles of government and private sector operation to solving the problems of water.

Given my familiarity with U.S. issues, it is natural that most of my examples come from them. However, I'm also able to tap international experiences from several sources, including work of colleague, the international community, and some of my own experiences, such as a project to study institutional arrangements and water law in Colombia.

The topics of governance are as broad as the subject of government itself, so the discussion is focused on those elements required for water management. Each process is explained in the context of the U.S. political system, which varies across a nation with 50 states and tens of thousands of local governments. Examples from other developed and developing nations are added to U.S. scenarios for comparative purposes.

REFERENCE

Grimond J. (2010). *For Want of a Drink*. The Economist. Special report. May 22, 2010, **395**(8683), 1–20.

Acronyms

(If not specified otherwise, organizations are of the United States)

7Q10	Ten-year low flow
AMPO	Association of Metropolitan Planning Organizations
ANA	National Water Agency (Brazil)
AWWA	American Water Works Association
BGD	Billion gallons per day
BMP	Best management practice
BNH	National Housing Bank (Brazil)
CAR	Regional Autonomous Corporation (Colombia)
CDPHE	Colorado Department of Public Health and Environment
CNPP&ID	Central Nebraska Public Power and Irrigation District
COG	Councils of Government
CVC	Regional Autonomous Corporation of the Cauca Valley (Colombia)
CWA	Clean Water Act
CWSRF	Clean Water State Revolving Fund
CWSS	Community Water System Study
DWSRF	Drinking Water State Revolving Fund
EBMUD	East Bay Municipal Utility District
EC	European Commission
EIS	Environmental Impact Statement
EMS	Environmental Management System
ESA	Endangered Species Act
FAO	Food and Agriculture Organization of the UN
FEMA	Federal Emergency Management Agency
FPA	Federal Power Act
GAO	Government Accountability Office
GEF	Global Environment Facility

GIS	Geographic Information System
GLAAS	Global Annual Assessment of Sanitation and Drinking-Water
Ha	Hectare
HAZUS-MH	Hazards U.S. Multi-Hazard
IBT	Interbasin transfer
ICPDR	International Commission for the Protection of the Danube River
ICPR	International Commission for the Protection of the Rhine
IID	Imperial Irrigation District
INBO	International Network of Basin Organizations
ISF	Instream flow
ISO	International Organization for Standardization
IWRM	Integrated Water Resources Management
LCR	Lead and Copper Rule
MAF	Million acre-feet
MCDA	Multi-criteria decision analysis
MHLW	Ministry of Health, Labor and Welfare
MPO	Metropolitan Planning Organization
MRC	Mississippi River Commission
MRC	Mekong River Commission
MS4s	Municipal Separate Storm Sewer Systems
MWD	Metropolitan Water District of Southern California
MWRI	Ministry of Water Resources and Irrigation (Egypt)
NARC	National Association of Regional Councils
NBI	Nile Basin Initiative
NCWCD	Northern Colorado Water Conservancy District
NEPA	National Environmental Policy Act
NIBS	National Institute of Building Sciences
NPDES	National Pollutant Discharge Elimination System
NWRC	National Water Research Center (Egypt)
O&M	Operations and maintenance
OFWAT	Water Services Regulation Authority (UK)
ORSANCO	Ohio River Valley Water Sanitation Commission
PUCs	State public utility commissions
RBO	River basin organization
SDWA	Safe Drinking Water Act
SFWMD	South Florida Water Management District
SRP	Salt River Project
SUDENE	Organization for Planning and Development in the Northeast (Brazil)
TBL	Triple bottom line
TBW	Tampa Bay Water
TMDL	Total Maximum Daily Load
TVA	Tennessee Valley Authority
TWM	Total Water Management
UCRC	Upper Colorado River Commission
UDFCD	Urban Drainage and Flood Control District
UN	United Nations

UNDP	UN Development Programme
UNESCO	United Nations Education, Scientific and Cultural Organization
USACE	U.S. Army Corps of Engineers (USACE)
USBR	U.S. Bureau of Reclamation
USEPA	Environmental Protection Agency
USGS	U.S. Geological Survey
WASH	Water and Sanitation for Health
WFD	European Union Water Framework Directive
WHO	World Health Organization
WRDA	Water Resources Development Act
WRPA	Water Resources Planning Act of 1965
WWTP	Wastewater treatment plant

Chapter 1

Governance and management for sustainable water services

Although the world is awash in water, only a tiny fraction of it is available as freshwater to society and the environment. Therefore, the possibility of a world water crisis is of deep international concern and vigilance is required to protect Earth's most valuable natural resource. The nature of the crisis will depend on your local situation, and it might involve water shortage and drought, rising sea levels, polluted water, floods, environmental degradation or other problems, but they all point to the same conclusion: water must be managed effectively, fairly and sustainably to support survival and prosperity for all people and living things.

Technical solutions for water problems are available and are improving all the time. The more serious problems lie in our governance and management systems. In a world with dramatically different situations and expectations, incentives work against sustainable water management. Although water law systems based on government regulation or economic incentives are in use, no system has been completely successful in addressing the many challenges of water management. What is required is a more complete and responsive system for water governance, one that includes water laws but goes much further to include the policies and support systems people need to manage water and care for the environment. That challenge forms the core topic of this book, which seeks to outline a responsive system for water governance that will apply to different problems in different places.

WATER GOVERNANCE AND SHARED RESPONSIBILITY

Water governance is about society's control levers that determine how the earth's most valuable natural resource is managed to meet human and environmental needs. It is the force behind water management, which sustains life support systems at many levels. It is a shared responsibility that spans the levels of government and the many uses to which water is applied.

In some ways, there is a fine line between water management and water governance. The distinction between them lies in the control function. Water management controls the resource to meet the needs of society and the environment. Water governance controls water management to make sure it does its job. If the two work perfectly together, it is possible to provide water to meet human and environmental needs at the same time on a sustainable basis (Figure 1.1).

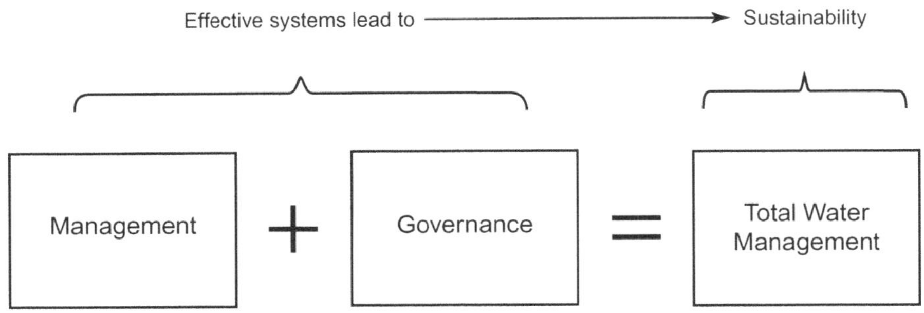

Figure 1.1 Management and governance for Total Water Management

Providing safe water for all and managing water for a sustainable environment are shared goals around the world, but they present daunting challenges. The only route to achievement of these goals is to apply effective water governance systems, which are the key to integration of policy, regulation, and capacity-building to coordinate the many competing purposes of water management.

Simply stated, a broad view of water governance is like parenting in a family. The parents make and enforce the rules, but they also provide the support the children need to succeed and comply with the rules. If they only focused on rules, what kind of family would it be?

Water governance helps sustain society and the environment by making and enforcing rules and providing support to water managers.

While it is one thing to have elaborate laws on the books, implementing them can be another story. Passing and enforcing laws are only the starting points to water governance, which must address broad issues that range from providing safe drinking water to protecting people from devastating floods, and do so in a spectrum of many cultural settings and countries. Given this broad variation in settings, effective water governance has high stakes and enormous challenges.

This book outlines a framework for water governance that should be valid in any political or economic setting. The framework offers processes for policy, empowerment, and control that can be adapted within different political, economic and cultural systems. The structure of the framework is comprehensive and can provide guidance for any water management scenario, and it offers choices for how the elements of policy, empowerment, and control can be applied.

The framework offers a way to govern water management that takes into account the world's many regions, cultures and levels of economic development. It also provides for integration across the purposes of water management, and it shows how to balance economic, social and environmental uses of water.

IMPORTANCE OF WATER GOVERNANCE

It is probably not necessary to explain the importance of water to the readers of this book. It has been my experience that once people become immersed in the subject of water, they become interested in how it is managed and how important it is.

Your perception of how critical water is depends on your situation. While many people have safe and reliable water services, billions of people still lack access to it and this leads to societal breakdowns, sickness, disorder, and environmental degradation. A range of examples can illustrate our shared deep dependence on water for life and prosperity and how water management and its governance apply to a range of situations. In rural Africa, you might have to walk long distances to fetch scarce but polluted water that will make you sick even while it is your only option. In Asia, you might experience a decline in fisheries or irrigation water that threatens your livelihood. If you live in a large western city, your problem may be caused by drought and your risk may be that your water utility depletes its reserves and imposes severe restrictions on water use.

While water governance is a critical input to sustainable societies, it needs improvement in many places. As the World Water Assessment Programme (2007) explained, in many countries water governance is in a state of confusion with a lack of water institutions or fragmentation of authorities and decision-making structures that result in many unsolved water problems that cause misery and poor living conditions.

"In many countries water governance is in a state of confusion"
– World Water Assessment Programme

Water governance can work towards solutions to these problems, but it cannot operate in an institutional vacuum. It requires help to address water needs that range across a broad spectrum, from survival requirements, such as access to safe drinking water, to discretionary water uses that raise the quality of life. In between are many economic, social, and environmental uses such as irrigation, energy production, and landscaping, among others.

Approaching water management as a shared responsibility is the only way to meet these needs sustainably because water systems are limited in capacity and balancing their many uses requires us to cooperate in managing common resources. This challenge is captured by the paradigm of the "Tragedy of the Commons," which explains why the public's resources are often neglected (see Chapter 8).

Managing water as a shared responsibility does not just happen, but it requires orchestration through systems that lie behind the scenes and are not as apparent as the front-line infrastructure and organizations that are required to meet water needs.

Water governance operates like the invisible atomic forces that hold elements and materials together. You may not see these forces but their effects are powerful and without them, everything would fall apart. In the same way, water governance provides both the empowerment and the control to keep divergent water interests from colliding with disastrous results. How the magic of water governance should work is introduced in the subsequent chapters.

Policy includes government plans and strategies at all levels. Control embodies all regulatory functions that apply to water, whether for public health and safety, environmental quality, performance or cost-of-service. Empowerment provides carrots along with control sticks, and it also offers a helping hand to individuals and organizations to overcome the inherent obstacles in managing water responsibly.

THE NEXUS BETWEEN WATER AND OTHER SECTORS

Water governance must be inter-sectoral to address its multiple purposes, apply across cultures and from low-income to high-income settings and have an institutional framework that stresses policy, empowerment, and control. This inter-sectoral attribute of water makes it unique among policy issues. In a practical sense, it creates linkages such as these:

- Water supply to poverty, health, and social policy
- Wastewater to housing, health, social policy, and environment
- Irrigation to food policy

- Flood to disaster and emergency management
- Hydroelectricity to energy
- Navigation to transportation
- Instream flows to environment

These linkages mean that you might be discussing water policy, but people in other sectors are discussing the same issues, but from a different slant. People see things differently. Say the topic is health of low income people in public housing. You see their safe water as a water issue, but other officials may see it as a housing and social issue. These different perspectives create the need for a lot of coordination.

The business of water is to handle a shared resource to meet needs across multiple sectors of society and nature.

A number of writers (including this one) have tried to create a model to explain the water sector, but we fall short because water defies a simple classification scheme. For example, Maxwell (2010) studied how water is organized and claimed that there is no such thing as a water industry, but instead it is a balkanized bazaar of quite different businesses, all of which have something to with delivery of clean water. As problems, he pointed to: untreated sewage, lack of access to safe drinking water and sanitation, water loss and decayed infrastructure, polluted rivers, endocrine disruptors, displaced people from new dams and obsolescence of old dams and, loss of wetlands, and depleted aquifers.

We can move past this dilemma by classifying the water sector as comprising organizations that actually handle water. The water-handling is the production side of the business, and it has also suppliers, customers and regulators, so it can be explained as a business.

By characterizing water as a business, you can explain its linkages to other sectors as producer-customer relationships. This is explained by a diagram of water sector links (Figure 1.2). In the diagram, you see how water and wastewater are linked to housing, how hydropower is linked to energy, and so on. Each of these linkages creates a nexus itself. For example, the water-energy nexus has been studied extensively, and you see many connections that go beyond the one shown for hydropower.

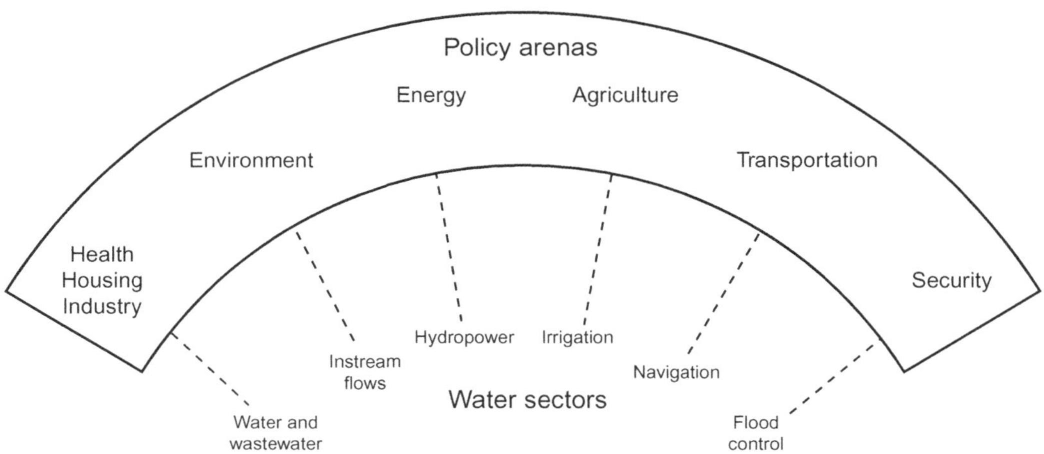

Figure 1.2 Water sector nexus links

GOVERNANCE FOR INTEGRATED WATER MANAGEMENT

To achieve success across the diverse purposes of water management, the inputs of effective governance, management systems, and infrastructure must be applied in scenarios involving different levels and sectors. Just as atomic forces hold diverse particles and elements together, the integrating force of water governance works with management and stakeholders to hold complex systems together. In particular, governance and management work together to pursue the goal of stewardship of water for the greatest good of society and environment, which is the definition of "Total Water Management" (TWM), as well as the concept behind Integrated Water Resources Management (IWRM) (Grigg, 2008).

Conceptual systems for integration in water management are needed from the global scale to the village level and for all uses, whether for economic, social, or environmental. The overarching goal of water governance is to enable these to work, but the difficult question is: "How to design water governance and make it effective?" To answer this question, we begin by probing what is to be governed and how the processes of water management should work.

To explain how the ideals of TWM and IWRM offer comprehensive frameworks for water management, you must drill down into the details of the physical assets, organizations, and controls that support the spectrum of water uses. These details are explained in Chapter 2.

How can the ideals of IWRM be met in spite of daunting political and social challenges? Clearly, technical inputs such as effective infrastructure systems are required, but without resources, good policies, and capable organizations and workforces, water management cannot succeed. The organizing principle of this book is to use the conceptual framework of water governance to show how these elements can be made to work. This conceptual framework is introduced in Chapter 3 and discussed for each water sector and issue in subsequent chapters.

Water governance links the inputs of resources, Rule of Law, and management to create the environment for sustainable water management (Figure 1.3). Resources represent a range of inputs that include finance, infrastructure and human resources capacity. The Rule of Law represents all institutions that are required for society to function successfully, such as a willingness of people to comply with laws and regulations. Management represents organizations and the application of knowledge to decisions required for sustainable water management.

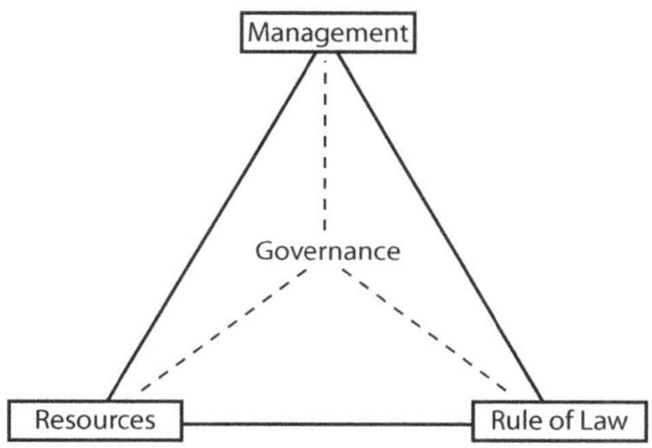

Figure 1.3 Governance to link requisites of sustainable water management

It is important to have a clear definition of water governance, but we must acknowledge that because it deals with complex issues, people will inevitably see it differently. The starting point is, "governance is the process of governing," but many questions about what this means must be answered and the subject goes a lot further than this simple statement.

Definitions of governance usually focus on administrative authority to manage public affairs at national, regional, and local levels. As an example, the Global Water Partnership defined governance as: "the range of political, social, economic and administrative systems that are in place to develop and manage water resources, and the delivery of water services, at different levels of society (Rogers and Hall, 2003).

The governance processes that apply most directly to water management are policy, empowerment, and control. While policy and control are recognized functions of government, empowerment is essential to build capacity so that all societies can meet water management objectives. Inclusion of empowerment as a legitimate function of water governance affirms that in today's complex societies, government must do more than simply control private activity. It also has a range of roles in meeting social needs and controlling environmental quality.

This broadening of the definition of governance is widely-recognized now. It was summed up this way: "Therefore, with regard to water, the focus of governance is the human and institutional resource capacities for the sustainable development and operation of water resources and management systems" (Osinde, 2008).

To see how policy, empowerment and control should work as the basic elements of water governance, imagine how they apply at the local level of government where most water management takes place. Here, governance is applied most directly by boards of directors (which could be called boards of governors). Imagine the situation if these local boards only thought they had a regulatory function and had no responsibility to help the water managers get resources or solve their organizational problems! This same logic applies at higher levels of government, so that at the national level, policy, empowerment, and control must be applied to enable success across-the-board for water management.

The regulatory programs of water governance respond to public concerns about safety, environment, and cost. Regulation of water allocation balances the allocation of water taken from natural systems. Regulation of public supply has a focus on public health. Regulation of discharges is the control of discharges into water bodies with a focus on health and environment.

Regulatory controls alone will not meet all water goals, and stewardship at all levels is required. That is, governance must go beyond top-down imposition of rules and controls to inspire individuals and organizations to care for water and its management systems and to work together with others in a shared approach to stewardship.

Special challenges to governance occur when water is managed in areawide, river basin, multipurpose, and trans-boundary settings. How these challenges require special coordination mechanisms and institutions is explained in Chapter 10.

A FRAMEWORK FOR WATER GOVERNANCE

Each element of the framework for governance has options, such as choices between the appropriation and riparian doctrines of water allocation law, regulatory versus economic incentive systems to control discharges of wastewater, or national versus state control of water permit systems. In other words, the required processes of water governance are generic and apply across-the-board, but cultural differences can be accommodated with choices among the elements of governance.

Two variables can be used to map the range of water governance issues across scales and intensity of problems. As shown in Figure 1.4, these are the level of government (political scale) and the condition of institutions (development level). By displaying scenarios on a map of these variables you can see the range of situations from water problems in a small village to national issues of a mega-scale developed country.

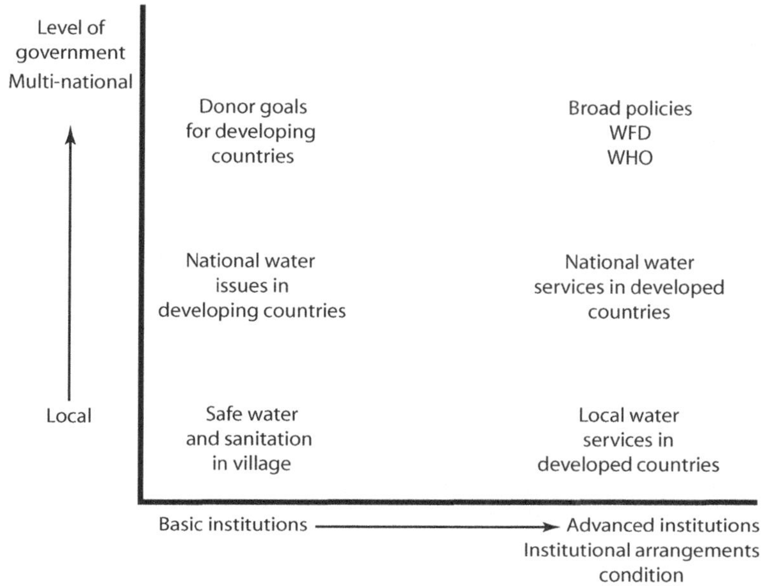

Figure 1.4 Water issues from small to large scale

THE SCOPE OF THE BOOK

This book presents on outline of the broad subject of water governance. It is meant to be short and to explain the framework of governance and set the stage for others to add substance about specific issues. The ideas in this book set a high bar, but they offer a spectrum of basic to advanced choices for the elements of water governance to achieve balance between water uses for different groups and between economic and environmental uses.

REFERENCES

Grigg N. S. (2008). *Total Water Management: Leadership Practices for a Sustainable Future.* American Water Works Association Press, Denver.

Maxwell S. (2010). A look at the challenges–and opportunities–in the world water market. *J. Am. Water Works Assoc.*, **102**(5), 104–116.

Osinde R. (2008). *Review of Existing Concepts of Water Governance and an Analysis of Pro-Poor Approaches to Un-Habitat Interventions.* United Nations Human Settlements Programme, Nairobi.

Rogers P. and Hall A. W. (2003). *Effective Water Governance.* TEC Background Paper No. 7. Global Water Partnership Committee, Stockholm.

World Water Assessment Programme (2006). *Water: A Shared Responsibility.* The United Nations World Water Development Report 2. UNESCO, Paris and Berghahn Books, New York.

Chapter 2

Water services management: purposes, functions and organizations

Water management requires the functioning of a range of participants in both government and the private sector. Neither is sufficient alone, but it is difficult to coordinate the functions in the wide band of participants and multiple levels. Responsibility for water in government is usually dispersed among different agencies. There is much interest today in water as a business, but if there is such a business it is not well defined.

People seem confused about water: is it a just commodity to sell, a natural resource to be preserved, or a unique public-private good that defies our attempts to manage it? We must answer this question to even begin a discussion of water governance. After all, just what is to be governed?

Water is a different kind of natural resource. A term like "renewable" distinguishes it from a "non-renewable" resource like oil, but then timber and fisheries are also renewable resources. How is water different from them? It is different by virtue of its continuous circulation through the hydrologic cycle and flows past one development after another. Water not only must be managed sustainably, it must be shared among many stakeholders with competing goals.

MODELING THE WATER SECTOR

The difficulty in managing water was explained in a special report on water in the Economist magazine: "No wonder a commodity with so many qualities, uses and associations has proved so difficult to organise" (Grimond, 2010). By "organise," the writer meant all aspects of water management and governance (Figure 2.1).

In the last chapter we saw how water management controls the resource and water governance controls management. Separating these two creates an artificial divide because sometimes management and governance converge. However, the dividing wall enables us to identify clearly the functions allocated to management and those allocated to governance. The chapter does this by presenting a model of the water sector and how it works when defined as a business. In the chapter we discuss how different types of water management organizations supply and manage water for the diverse and often-competing purposes that it serves.

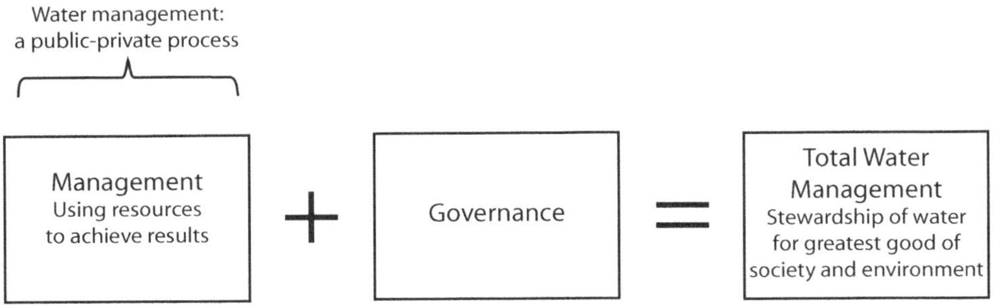

Figure 2.1 Management functions and governance for Total Water Management

Although water governance and management often converge, they have distinct roles in meeting water needs.

THE WATER SECTOR AND ITS SUBSECTORS

The starting point for explaining the sector is to outline what it includes. For terminology, we use the "water sector" to refer to all water services (or subsectors) lumped together. The water sector is sometimes called the water industry, but in the last chapter, we cited Maxwell's (2010) claim that there is no such thing as a water industry but instead a bazaar of quite different businesses, all of which have something to with delivery of clean water.

Our view is that the water sector involves more than just delivery of clean water as a commodity and that it has attributes of an industry, but must be approached as a complex, multi-faceted sector. In other words, water management may be a cross-sector activity, but it is necessary to draw boundaries around it and explain it because it has unitary characteristics that focus on the management of the resource.

The water sector is difficult to classify because of this complexity, but people working in it have an intrinsic understanding that its core players are the organizations that handle water. Following that conclusion, it can be explained as a business with producers, suppliers, customers, and regulators. In that sense, it is like an industry and it is a sector of the economy, broadly defined.

The business of water has producers, suppliers, customers, and regulators.

As shown in Figure 2.2, the sector can be mapped onto a series of concentric rings with the action held together by coordination and regulation from the center. In the first ring, the water-handlers provide services by managing water, including water supply, wastewater, stormwater, irrigation and drainage, dams and reservoirs, groundwater, instream flows, and flood control.

The next ring shows the categories of water customers, which include: residential users, major industrial and energy water users, businesses and government users, farming, land development, landscaping and erosion control, the environment, fish and wildlife, recreation, hydropower, and navigation. The outer ring of the model shows the suppliers to the sector, including equipment, materials, construction, and professional services.

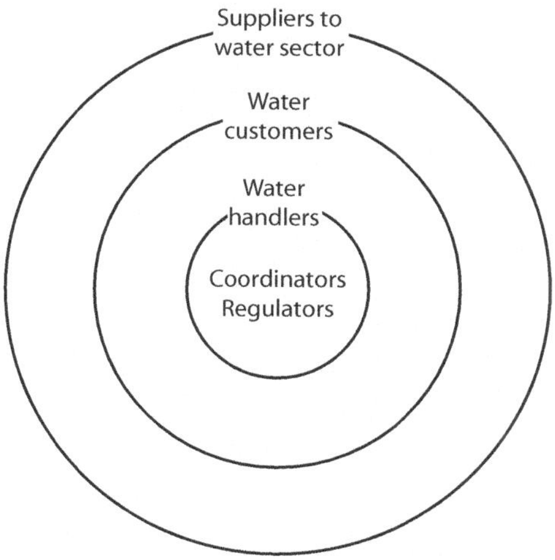

Figure 2.2 Water sector elements

Regulators control water allocation and use, water quality in the environment, public water systems safety, land use in flood plains (floodplain management), and emergency preparedness for droughts and other water emergencies. In some cases, we refer to these as subsectors of the overall water sector so as to delineate regulatory functions as opposed to management functions.

CLASSIFYING WATER SERVICES

To identify how water governance should work, you must explain the functional elements of the water service categories. For example, urban water supply is a service to customers, which requires the allocation of raw water from environmental water sources. Here you have two functional elements (providing treated water to customers and regulating the allocation of raw water), which involve different constituencies, regulations, and government agencies. Thus, governance processes for providing service and allocating raw water must be disaggregated to analyze how they should work.

By the same token, management of wastewater involves different agencies than regulation of stream water quality. Therefore, water quality regulation is treated as a separate sector from wastewater management and requires different governance elements.

The nature of the required water governance will depend on the following variables:

- The specific service or commodity delivered
- The extent to which the service benefits individuals or the public-at-large
- Whether the organizations are in the public or private sector
- The level of government involved
- Whether the organizations deliver or regulate services
- Their financial and organizational capacity

For example, if services benefit individuals or the public-at-large determines whether they can be sold on a self-supporting basis as commodity goods or whether subsidies are required. If service organizations are in the private sector, they are usually regulated for business affairs by government agencies, but if they are in the public sector, they are probably self-regulated. Both types might be regulated by higher level government agencies for health and safety.

Whether the organizations deliver or regulate services determines how they relate to each other. The regulators should be separated from the service providers. In some cases, however, this is not practical because the service providers may lack the capacity to meet the regulations. Examples will be discussed later in the book. In this case, financial and organizational capacity may require assistance from government agencies.

Let's consider first the services (Figure 2.3) that mostly benefit individuals and organizations and which can mostly be sold as commodity goods. These include water supply, wastewater, and irrigation water. Stormwater services, which are shown with a dashed line to indicate that they are a different type of service, can sometimes be included with this group, as will be explained in Chapter 9.

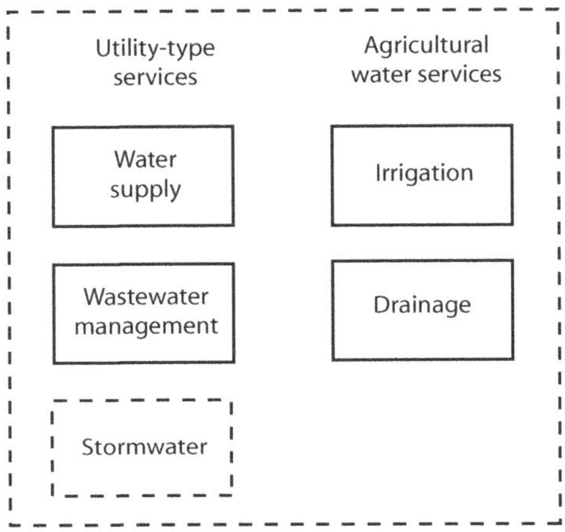

Figure 2.3 Water services for individuals and organizations

Next, we look at the water services and controls with mostly public purposes (Figure 2.4). These include planning, coordination, and regional agencies; instream and groundwater management functions; regulation; and protective services. Regulators are also shown to illustrate that they provide services too. Water allocation regulators control water supplies and the water quality category includes both public water supply regulation (with its focus on public health) and environmental water quality (with its focus on natural water systems).

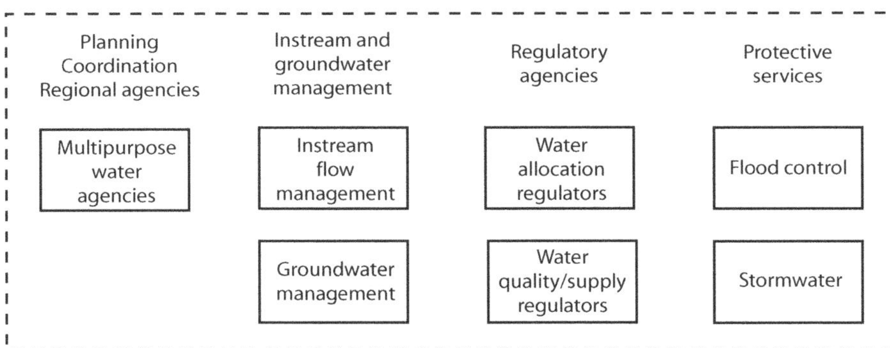

Figure 2.4 Water services and controls for the public and groups

By distinguishing the services that benefit individuals from those that benefit the public-at-large you can discern differences in governance requirements and opportunities. If you also consider the differences between large, high-capacity organizations and smaller, lower-capacity organizations, you can see how the need for governance ranges from regulating leading-edge utilities to helping small and struggling community-based programs. As examples, consider that in the U.S. water supply utilities include some 53,000 community water systems, most of which are very small, and irrigation organizations range from large well-financed districts to small irrigation associations that serve only a few farms.

WATER SERVICES AND REGULATION: A WATER CYCLE MAP

Regardless of the type of water service, you can map it to show its effects on the water cycle (Figure 2.5). It is worth the time to explore the diagram to see how the services, management actions, and regulatory controls relate to each other.

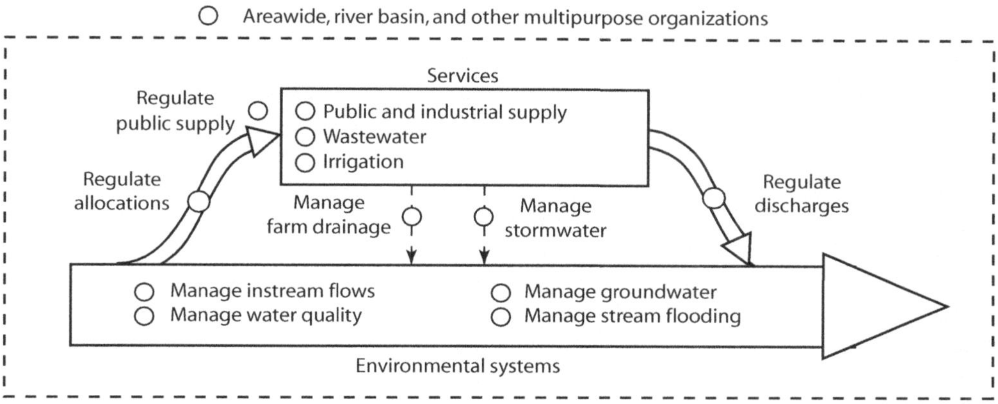

Figure 2.5 Water management and regulation functions

Figure 2.5 shows 13 separate water management functions, each with its own purposes and controls. The functions may seem similar, but they are distinct and fit into different categories to provide services directly to customers, to manage water in the environment, or to manage residual water that returns to it. One function coordinates water management and services, and three others regulate water uses and management.

To trace how these functions interact in water management, consider the services of water supply and irrigation, both of which furnish water as a commodity. Public and industrial supplies are provided by utilities and self-supply facilities for drinking and other purposes. Irrigation is the supply of raw water provided by organizations and individuals to serve farms, ranches and aquaculture. Taken together, public and industrial supply and irrigation represent the major uses of water for economic and social purposes.

To understand these water supply categories, a number of water use categories must be considered to prepare a complete accounting of water use. Examples such as water for energy, aquaculture, livestock, and water losses belong to the categories of public and industrial supply and irrigation.

It follows that when water is supplied, it generates the need for management of residual wastewaters. In addition to piped wastewater from residences and businesses, other water residuals include drainage from land uses (stormwater) and from farming (farm drainage). Wastewater service supports sanitation services by collecting residual water and removing it from buildings and points of water use. Farm drainage is for removal of drainage water from farms in irrigated or rain-fed areas and can be accomplished by individuals or organizations. Stormwater programs handle runoff and minor flooding from land use sites, as well as control of the quality of nonpoint runoff. While these residuals occur from different activities, they blend together into a waste stream that requires similar water quality governance and management actions.

The functions of environmental water management are for control of the "water commons" in the sense that water in the environment, whether surface water or groundwater, flows from one place to another and is used successively for diverse purposes. Given this shared use, the problem of the "Tragedy of the Commons" must be faced in managing environmental water. This slogan, which is explained further in Chapter 8, refers to the idea that when resources belong to everyone they will be abused. Only when they belong to individuals as property will they be managed well.

The categories of environmental water shown on Figure 2.5 are instream flows (quantity), water quality, groundwater, and stream flooding. Instream flows are for hydropower, navigation, recreation, fish and wildlife, and to convey water to users. Water quality is for regulation of instream water quality, including permits, monitoring, and enforcement.

Groundwater management is for regulation of groundwater withdrawals and actions such as aquifer-storage-recovery projects. Stream flooding includes structural and nonstructural programs to protect people and property from damages and to manage floodplains. These may seem too disparate to categorize together, but the logic is that they comprise the movement of water on the surface and underground, they address both water quantity and quality, and they include flows during normal times and during flooding. As we continue the discussion of environmental water, how these fit together in governance and management will be clear.

The category of areawide, river basin, and multipurpose management captures the function of coordination across geographic areas and purposes of water management. It is necessary to deal with the water commons issue and to bridge the gaps among other competing uses. The management agencies meet coordination and other needs, such as direct water services and/or river basin planning.

Of the functional areas described above, infrastructure systems are required for supply services (water supply and irrigation); managing residuals (wastewater, stormwater and farm drainage); and flood protection (for structural approaches). Functions that involve mostly program management are environmental water management (instream flows, water quality, groundwater, and floodplain management); areawide, river basin, or multipurpose organization management; and regulatory programs (allocation, public supply and discharges). In some cases, areawide, river basin, or multipurpose organization management may also involve water management infrastructure.

The regulatory programs to control water allocation, public supply and discharges represent a broad set of public concerns about safety, environment, and cost. Regulation of water allocation is the control that balances the allocation of water taken from natural systems. Regulation of public supply is the control of the quality of water provided by public systems and has a focus on public health. Regulation of discharges is the control of discharges into water bodies with a focus on health and environment.

A TASK MAP OF WATER MANAGEMENT

Referring again to Figure 2.5, we ask now: "What are the management tasks that create and unify the water industry?" Descriptive terms such as planning, controlling, and operating begin to tell the story, but they are general terms and must be applied to specific examples to provide clear explanations.

Of course, any water service organization will execute myriad management tasks to achieve its mission. It will employ its infrastructure and engage its workforce to plan, organize, and control operations. As in other businesses, it will also undertake support tasks such as legal, human resources, and purchasing. The management tasks of concern here are those that affect the delivery of water services in a direct way, and these focus on operations and capital decisions.

Operating decisions determine how an organization's infrastructure and other resources are used to manage water. Examples are to operate a reservoir, a treatment plant, a distribution system, an irrigation system, a monitoring system, or a program for risk management. Operating results determine the quality of service and will come under regulatory control for most water services, such as drinking water supply.

Capital decisions are required for the infrastructure and equipment necessary to fulfill the purposes of water management. Examples include: purchasing water rights, building a wastewater treatment plant, and rehabilitating a dam for safety purposes. These decisions also come under regulatory control because infrastructure adequacy is an important determinant of operating performance.

Financial decisions determine whether the organization has the capacity to build and operate its facilities and programs. Examples include programming and budgeting, requesting a rate increase, and deciding on a bond issue. Government-owned utilities and water organizations are mostly self-regulated and subject to political control. In cases of privately-owned water companies, regulatory commissions may have authority to approve financial decisions such as rate increases.

Regulatory decisions ensure compliance with the law and rules established by government. Examples include setting a rule, issuing a permit, allocating or rationing uses, taking an enforcement action, operating a regulatory monitoring program, or regulating a flood plain. Regulators must be controlled to ensure that the regulatory process is fair and balanced. Regulatory controls are illustrated by Table 2.1.

Table 2.1 Regulatory controls

Regulatory category and goal	Example
Quantity allocation	Balance allocation of water rights
Service quality	Provide water at adequate pressure and flow
Business affairs	Control fees and infrastructure condition
Health and safety	Supply drinking water that is safe
Environmental water quality	Maintain fishable and swimmable streams
Environmental water quantity	Maintain instream flows for habitat

WATER MANAGEMENT ORGANIZATIONS

Water management organizations fall into the general categories shown on Figure 2.5. The types of organizations and what they do are explained in the later chapters about the water management sectors. Table 2.2 gives a few examples of organizations in each category (note that some of them offer multiple services and fit into more than one category).

WATER MANAGEMENT SUCCESS FACTORS

Regardless of the situation, effective water management requires a combination of technical and institutional factors that include infrastructure, financing, and management capacity. Institutional factors include basic arrangements such as the legal system, organized governments, and supportive socio-cultural systems. Without effective institutional arrangements, water management will not succeed well even if technical factors are in place.

Table 2.2 Types of organization (by services provided)

Water supply	Denver (Colorado) Water, Tokyo (Japan) Bureau of Waterworks, Sydney Water (Australia)
Irrigation and drainage	Imperial Irrigation District (California), Jordan Valley Authority (Jordan)
Wastewater	Metropolitan Water Reclamation District of Greater Chicago (Illinois), Acueducto de Bogotá (Water and Wastewater Authority of Bogotá (Colombia)
Stormwater	Urban Drainage and Flood Control District (Denver, Colorado), Visakhapatnam Urban Development Authority (India, multipurpose authority)
Instream flows	Saskatchewan Watershed Authority (Canada) (Note: this is an example of a comprehensive watershed authority, as an authority only for instream flows would be rare)
Water quality	Ruhr River Association (Germany), Ohio River Sanitation Commission (US)
Groundwater	Edwards Aquifer Authority (Texas), Central Ground Water Board, Ministry of Water Resources, (India)
Flood control	Miami Conservancy District (Ohio), Rijkswaterstaat (Netherlands National Water Agency)
Areawide, river basin, and multipurpose management	Lake Constance Water Supply Association (Germany, water supply), South Florida Water Management District (Florida), Regional Corporation of Cauca River Basin (Colombia)
Regulatory water allocation	Division of Water Resources (Colorado), Regulatory Agency for Water, Wastewater, and Energy of the State of São Paulo (Brazil)
Regulate public supply	Colorado Department of Public Health and Environment (US), Water Services Regulation Authority (Ofwat, UK)
Regulate discharges	New York State Department of Environmental Conservation (US), Ministry of the Environment (Finland)

Effective water management involves both technical and institutional factors.

Institutional arrangements range across formal laws and organizations to informal arrangements of cultural behaviors and how people work together to solve problems. Governance draws on institutional arrangements to control water management outcomes. The biggest obstacle to success in water management lies in overcoming conflicts over values, property, and political objectives. The institutional arrangements to address these and other cultural and behavioral issues are the key to success and governance is the mechanism to address them. Without effective governance, water management cannot meet the often-competing needs of economies, social systems, and environmental systems.

WATER MANAGEMENT CHALLENGES

The main governance challenge is to integrate water resources management to pull together the many conflicting forces, which are often reported with scary headlines. How valid are these headlines? It is well to consider them in concluding this chapter with a discussion of water management challenges.

Returning to the special report in the Economist magazine, we can parse the topics to discuss water management challenges (Grimond, 2010). First, there are a lot of scary headlines, but most analysts conclude that problems are manageable if we can get our act together. The Economist article included this warning: "Everyone must use less water if famine, pestilence and mass migration are not to sweep the globe." It acknowledged that the language is sometimes overblown, but the message in not incorrect.

Then, there is the question of drivers. What is causing the problems? Answer: rising population and living standards. Overuse of water results in falling aquifers in megacities like Mexico City and in cities like Jakarta and Buenos Aires the overdrafts are accompanied by subsidence, pollution, and salinization. Agricultural aquifers are under stress too.

The supply is limited and there are a lot of hydrologic issues, like the overdrawn aquifers, overused rivers, and pollution. Climate change is a new threat. In addition to farms and cities, industry takes a lot of water too. Then, we have the problem that people don't expect to pay full cost for what it costs to manage water.

But the problems are spotty. Some countries, like India, take enormous quantities for irrigation. Others, like Britain, take little. The spottiness also occurs within countries and over the span of an average year.

All of this results in water scarcity, both hydrologic and economic as well as big environmental problems. It is much worse in some countries than others, and in some problem countries, the population is growing fastest.

The water issues involve a lot of fuzzy terminology and complex economics, such as whether water is a public or private good. It is not surprising that the Economist writer concluded: "No wonder a commodity with so many qualities, uses and associations has proved so difficult to organise."

Because water is considered as both a public and a private good, its management poses special challenges to the economic process of allocating society's resources. Water offers many beneficial uses that include supplies for various purposes, the capacity to assimilate pollutants, environmental nourishment, and others. Also, using water-related resources, such as reservoir or channel capacity, provides security against damage from flooding.

Concepts of water management range from simple cases—such as to control flow in a pipe—to complex situations involving large-scale political problems. In searching for a way to explain the need for a comprehensive approach, writers turn to complex concepts like Integrated Water Resources Management and Total Water Management). While these are difficult to explain and involve many elements, at their core they explain the need for balance in water decisions (Grigg, 2008).

Ideally, water management would work through voluntary cooperation among water users, but the competing agendas of water users lead to conflicts that require involuntary rules, regulations, controls and incentives. These become elements of water governance, which is required to steer the overall process of water management.

While this is just a snapshot of water issues, it shows that this sound bite applies to water: "We have met the enemy and the enemy is us." The problems are manageable, but we have to get to it.

"No wonder a commodity with so many qualities, uses and associations has proved so difficult to organise" – Economist magazine special report on water

FROM MANAGEMENT TO GOVERNANCE

Each of the management functions, tasks, organizations or issues requires its own set of governance actions. The next chapter explains the general elements of governance, and then sector issues are explained in a series of chapters. Some of the governance issues to be highlighted in these forthcoming chapters are shown in Table 2.3.

Table 2.3 Governance Issues

Water allocation	Equitable water allocation is one of the principal challenges to address conflicting uses, including the environment and allocating water during shortages. Governance to achieve water use efficiency is an ongoing challenge.
Water supply	Water supply involves challenges from access to safe water to all to sustaining reliable service in millions of small systems globally, and managing large centralized urban systems with their aging buried infrastructures.
Wastewater and water quality	Access to sanitation is required for public health. Managing millions of small and individual systems, as well as large wastewater collection and treatment systems with their aging infrastructures is required for both health and environment, and regulating industrial and municipal discharges, along with nonpoint sources is a challenge.
Irrigation and drainage	Managing large volumes of water use across ranges of organizations and individual systems and mitigating harmful effects of drainage waters present large challenges that are linked to the food sector.
Instream flows	Many people depend on coordination of shared uses in the water commons, and working out the conflicts is a challenge to governance.
Flood control and stormwater	Dealing with the cumulative effects of development and solving inter-jurisdictional drainage problems are challenges, along with mitigating threats to vulnerable people living near flood hazard zones.
All systems	Many challenges involve more than one water service or links with other policy sectors. These include providing safe and sustainable plumbing systems and services globally at the household level and creating healthy living conditions to combat the range of water-related diseases. Sustaining shared approaches to governance at scales from local to global and building capacity with a sense of responsibility and stewardship are required. Pricing mechanisms are needed to achieve efficient water allocation and services.

CONCLUSIONS

Our goal in this chapter was to present a model of the water sector and explain how the subsectors work as businesses that serve the diverse and often-competing purposes of water uses.

It is the role of managers in the water subsectors to serve these purposes and the role of governance to balance and control water use for the "greatest good" of society and the environment. They perform their jobs by planning, organizing, and operating the infrastructure systems required to meet water purposes, and these include tasks ranging from complex river basin planning to individual projects like pumping stations, for example.

The infrastructure systems they manage can be massive and capital-intensive. They are required for the water supply services, to manage residual waters and to protect against floods. Water managers also operate programs that do not involve as much infrastructure. These include instream flows, water quality, groundwater, and floodplain management. Also, areawide, river basin, or multipurpose organizations manage mixtures of these services, and regulatory programs operate alongside all other water subsectors to control allocation, public supply and discharges.

In the context of these water management elements, it is the role of governance to enable and control the organizations that must operate effective and sustainable water systems and services. The processes of governance extend across water management sectors and form a nested set of activities that empower and regulate the subsectors, and coordinating management and governance is the key to fulfilling ideals such as Total Water Management or Integrated Water Resources Management.

Water management and governance differ among countries. In the developed countries, the needs are tuning up, improvement, and sustaining water management but we cannot afford to be complacent or to think that all problems have been solved. Challenges include diligence in providing water for environmental systems and instream flows, vigilance in regulation of drinking water and wastewater, gaining control of nonpoint sources especially stormwater, improving allocation of water for social equity, empowerment of IWRM organizations, and dealing with aging infrastructure. These challenges must be faced in the context of financial limitations, along with a host of political and other institutional issues.

In the middle and low income developing countries, issues range from the most urgent (access to water and sanitation) to development of infrastructure (such as for wastewater treatment), to irrigation assurance in some countries, to protection of people from flooding (in cities as well as rural areas), and to gaining control of environmental issues when they are able to deal with basic issues. Institutional capacity-building is the biggest step needed in almost all cases.

Postscript

One of the pitfalls of working in a mostly-public business like water is that you might under-value the contributions of private enterprise. I've seen this while working with one foot in the public sectors of government and academia and the other in the private sector. I formulated a joke (which I'll spare you from) about it to emphasize the tendency. The joke showed how little appreciation that people in government, business and academia had for the other communities.

So, in recognition of the desirable qualities of government and business I and a few fellow travelers have made attempts to define water as a business. This can spin out of control quickly and become cumbersome. We tend to over-organize things and by the time you have classified all water management processes you end up pretty quickly with too much complexity.

I've also learned that this complexity fascinates economists, and if you look into the subfield of public choice economics you find many interesting studies of how to jointly organize public and private sector activities such as water management. My take on it is to define water management and the water business as I did in this chapter, by focusing on the actual handling of water as a resource.

At the end of the day, we can't avoid the complexity, however, and concepts like IWRM and TWM have been developed to provide frameworks for it. My book on TWM (Grigg, 2008) explains my aspects of it, but I'd be the first to admit that we have a lot of work to do to make water management more businesslike, while considering fully its Triple Bottom Line effects (see Chapter 11).

REFERENCES

Grigg N. S. (2008). *Total Water Management: Leadership Practices for a Sustainable Future.* American Water Works Association Press, Denver.

Grimond J. (2010). For want of a drink: a special report on water. *Econ. Mag.*, **395**(8693), 1–20.

Maxwell S. (2010). A look at the challenges–and opportunities–in the world water market. *J. Am. Water Works Assoc.*, **102**(5), 104–116.

Chapter 3

Water governance *framework and principles*

Water governance is essential to a sustainable future, but it sounds complicated. What is it, anyway? Scholars debate its many attributes and it sometimes seems you need an advanced degree in the social sciences to follow the debate. Clearly, it involves institutional issues, but what are those? Some scholars think the term institutional should be discarded because no one really understands it.

You can't avoid the reality that water governance is complex because it is connected to the fundamental systems of government and politics of a country. The many conflicting agendas of water users run against each other and seem to block the kind of cooperative management that is needed. In spite of these challenges, water governance can be explained in the context of a few core processes that are necessary at each level of its application. In a basic sense, these are policy-setting, capacity-building, and regulating water management.

WATER GOVERNANCE AND WATER MANAGEMENT

Although water managers may deliver the services required, they encounter many obstacles, especially in achieving integrated solutions that balance the resources and sustain environmental needs. The obstacles are often called "non-technical" because they cannot be solved by the application of engineering. They involve inadequate resources and capacity and conflicts over values, property, and political objectives. Without a working governance system, water managers cannot overcome these obstacles.

To overcome them, governance systems should act in parallel with management to work out issues of policy, empowerment and control of water systems (Figure 3.1). Management and governance are different, but both have essential and inter-related functions in integrated water management. The success of governance will be driven by institutional arrangements that are explained in this chapter.

The chapter presents a framework for water governance and explains the principles that guide policy, empowerment, and control of water management organizations. It also provides examples of the decisions that require governance actions, and it explains these decisions across the ranges of water services and types of countries.

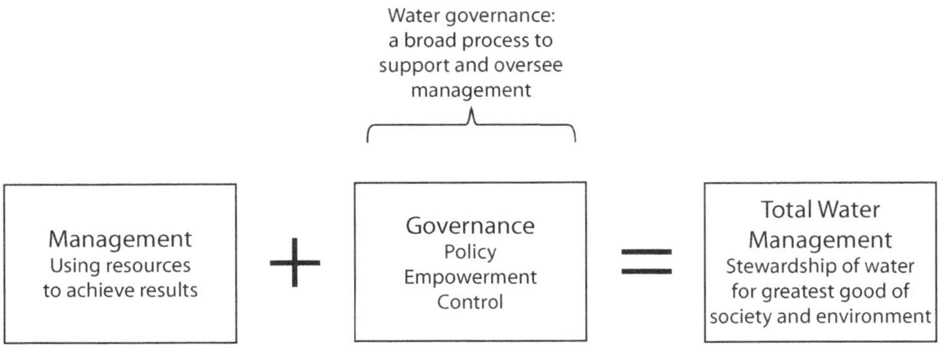

Figure 3.1 Management plus governance elements

HOW DO GOVERNANCE AND MANAGEMENT RELATE?

Doesn't management also deal with policy, empowerment and control? If so, how do these functions compare between management and governance? The answer is that policy and control are abstract concepts and they only become meaningful when they are applied to specific situations.

Governance focuses on control of all activities in a sector and is a higher-level process than management, which operates at the level of the service organization (Figure 3.2). Although their functions sound alike (such as policy and financing), the focal points of management and governance are different. Management's responsibilities cut across the spectrum of organizational policy, planning, directing, and controlling, while governance responsibilities are focused on sector policy, empowerment, and regulatory control. In other words, you must look past the abstract language to see the nuances between the functions of management and governance.

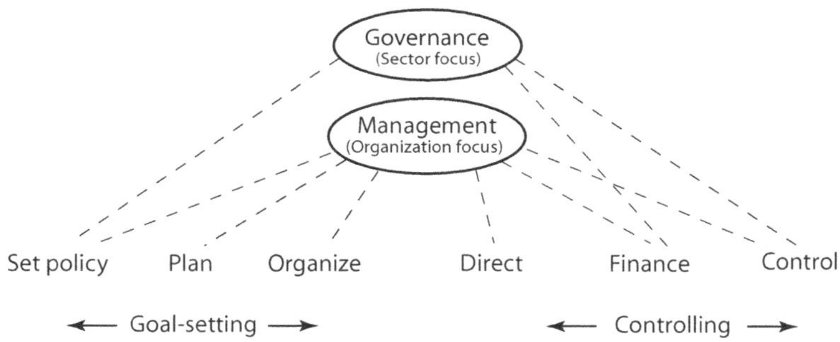

Figure 3.2 Functions of water governance and management

The distinctions between management and governance may seem like fine points, but they explain why managers and regulators fight rather than cooperate as desired in IWRM. Management's responsibility is to use its resources to accomplish organizational missions, while governance guides, empowers, and controls actions for groups of organizations. Managers and governance officials have different incentives and operate with a certain amount of tension, especially in the regulatory process.

The division of responsibility of governance and management varies across countries because how governance is applied depends on the context. However, their combined effectiveness largely determines the success and sustainability of water management and services in any situation.

DEFINITION OF WATER GOVERNANCE

It is common to define governance as the process of governing, but this simple phrase does not explain it fully because governing itself is complex. At the end of the day, water governance is about doing what it takes to enable the sector to succeed. The broadening of the concept was explained in Chapter 1.

At one extreme, governance can be viewed as government control, mainly through the regulatory process. However, a broader view expands its roles to address the totality of instruments and mechanisms that influence social change (Lafferty, 2004). The UNDP (1997) definition of governance seems to provide a middle ground in explaining it: "The exercise of economic, political, and administrative authority to manage a country's affairs at all levels. It comprises the mechanisms, the processes and institutions through which citizens and groups articulate their legal rights, meet their obligations and mediate their differences."

Governance is the "... exercise of economic, political, and administrative authority to manage a country's affairs at all levels." – UNDP

Keep in mind that governance operates at all levels as well as at the country level. The board of governors of a small water department is engaged in governance, even if its span of responsibility is limited.

The political challenge of water governance is found in this question: Where do you draw the line between public responsibilities and private rights in controlling the use of water resources? There is no single answer, of course, and the answers must be worked out in case-by-case approaches to the many different types of water problems found around the world.

The design of policy, empowerment, and control depends on the level of government and on the capacity of the organization delivering water services. At high levels, such as the United States or the European Union, water policy will address large-scale issues and details of water management will be left to lower levels of government. For example, the European Union Water Framework Directive (WFD) provides guidance for water governance in member states (Barreira, 2006). At the lower level of individual utilities, the corporate governance must take into account policies set at higher levels and add its own more detailed policies and programs to ensure successful outcomes.

The capacity of the organization delivering water services will determine how governance is applied because a high-capacity organization can normally take care of its own resource and empowerment issues and require mostly regulatory control. However, if the organization is in a low-income country and lacks resources and capacity, governance will need to help it along, rather than simply regulate it.

INSTITUTIONAL FRAMEWORK FOR WATER GOVERNANCE

The elements of governance required depend on the water service, the benefits to individuals versus the public-at-large, the public–private mix, the level of government, and the function and capacity of the organizations being governed. No matter how these are aligned, governance will be based on a chain of institutional arrangements such as those shown on Figure 3.3, which illustrates how its foundation comprises fundamental institutions, such as the Rule of Law. Then, institutional mechanisms such as enabling legislation are required. These must be implemented by organizations and structures, which work through strategies such as plans and policies to achieve desired outcomes.

It would be well to stay with the simple diagram shown in Figure 3.3, but the framework for water governance must be expanded to explain it more completely. Figure 3.4 is a map of the arrangements needed to make water governance work. While you cannot ignore the complexity shown on the diagram, it does not have to be absorbed at once because it is rather like a broad picture of the political science of water. To be even broader, you would have to expand the informal mechanisms, which are shown at the bottom of the diagram.

Figure 3.3 Framework for water governance

The map shown on Figure 3.4 illustrates the institutional arrangements that are required to make water governance work. The expression "institutional arrangements" is familiar to social scientists but not always to scientists, engineers and managers. It is a powerful concept, although hard to explain because it involves such complex and interwoven issues. Basically, institutional arrangements determine how society evolves through the rules and norms of interaction (North, 1990). These rules involve the established organizations, processes, and legal arrangements of society, and the norms determine the societal forces that determine how we do business.

Having disclosed that the institutional map involves many complex elements, let's move around it some. It shows that governance begins with top-down controls on society that are based on the Rule of Law and organized government. You might think of U.S. law, for example, as having evolved from European political and legal systems, mostly from Britain, from where our early water laws were derived.

At the bottom of the diagram, you see how governance includes informal institutions such as incentive structures that guide our actions. This explains realities such as we see in our own local water law, where some insiders say: "there are three ways to view Colorado water law—the way it is written, the way people think it works, and the way it really works!" In other words, you must be involved in detailed interpersonal transactions to really know what is happening.

From the diagram, you can see how policy, empowerment, and control depend on these institutional arrangements. Some of the important canons that it illustrates are:

- The Rule of Law and organized government are essential for a successful society and water governance will not succeed unless they are in place and effective
- Basic institutional mechanisms to facilitate water governance include government authority, contracts and agreements, enabling legislation, and regulatory capacity
- Water management, regulatory, and empowering organizations are required to implement the roles and processes of water management

- Water policy and strategy are derived from government actions and public participation
- In addition to these, water management is influenced by the informal forces, networks, and patterns of behavior shown at the bottom of the diagram

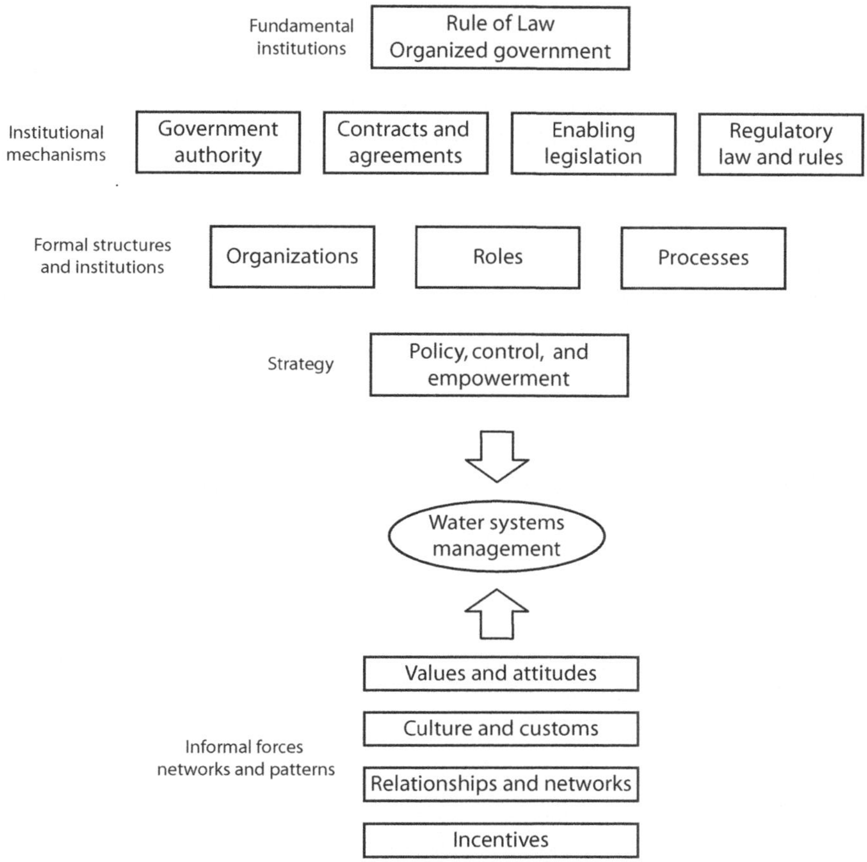

Figure 3.4 Map of institutional arrangements for water governance

This framework of institutions shows what is needed, but they must be mobilized. I credit my Colorado State University colleague Evan Vlachos with introducing me to the term "institutional mobilization," which explains well how it is necessary to go beyond having institutions to the point where they work effectively.

Now let's identify some principles of water governance that flow from the institutional arrangements. How these institutional concepts translate into water governance is explained in the subsequent chapters for each sector.

PRINCIPLES OF WATER GOVERNANCE

How you derive principles of water governance depends on which water services you include and what you mean by governance. For example, in a book on institutional governance and regulation for the water sector, Rouse (2007) started from the UNDP definition and derived the eight principles of governance shown in Table 3.1.

Table 3.1 The eight principles of governance according to Rouse (2007)

Government organization	Government structures in which sound integrated policies can be developed
Government capacity	The professional capability within governments and civil services to develop sound policies
Separation of powers	Separation of policy, regulation, and delivery
Independent regulators	Regulatory bodies that are independent and that have the power to implement policies, laws and rules in a transparent way
Monitoring and enforcement	Effective monitoring and enforcement of regulations
Transparent operations	Open and transparent information and reporting as the most effective means of combating corruption
Public participation	Effective public participation as part of transparency, but also to provide for buy-in and commitment
Empowered management	Utility operators who, having been given clear objectives, targets and resources, are free from political interference, so that good operational management can concentrate on providing efficient services

Looking at the left side of the Table 3.1, you see that these are really broad concepts that apply to a number of government issues and sectors. You might say that they illustrate that governance for water is the same as other sectors because it depends on government doing its job well. That is certainly true, but it is not the end of the story. Water governance involves many more details and nuances that apply specifically to its unique attributes and scenarios.

The principles in the list illustrate the importance of the governance elements we are discussing, policy, empowerment, and control. Policy is illustrated by the principles of government capacity and the need for professional capability within governments and civil services to develop sound policies, as well as by the principle of government organization to have structures in which sound integrated policies can be developed.

Empowerment is embedded in these principles, but is not as explicit as policy and control. It is illustrated by empowered management and utility operators who, having been given clear objectives, targets and resources, are free from political interference, so that good operational management can concentrate on providing efficient services. Also, it is illustrated by the principle of public participation and the need for transparency and to provide for buy-in and commitment.

The governance element that is most apparent in the principles is control, as shown by the following principles: separation of powers for policy, regulation, and delivery; regulators that are independent and that have the power to implement policies, laws and rules in a transparent way; effective monitoring and enforcement of regulations; and transparent operations to provide open and transparent information and reporting as the most effective means of combating corruption.

Rouse's focus was on water and wastewater. To expand the discussion about policy and empowerment across the range of water services, we can focus on water management issues such as the need to empower small utilities and enterprises, corporate governance, and stewardship. This can enrich the discussion of utility management and regulation and increase our understanding of integrated water management issues.

POLICY, EMPOWERMENT, AND CONTROL IN WATER GOVERNANCE

From the institutional map of Figure 3.4 it is a logical step to move to a discussion of the use of policy, empowerment, and control to influence the performance of water management systems. These topics are also discussed in detail in subsequent chapters: water policy in Chapter 17; empowerment in Chapter 12; regulatory controls and water law in Chapters 15 and 16; and control through planning and coordination in Chapter 17. Many examples of these topics are also included in the tool box of the Global Water Partnership (2010).

Water policy and strategy

Policy-setting is the highest-level governance process because it establishes the goals and objectives for water services. As an example, goal-setting is the policy process of setting targets, such as "safe water for everyone."

Water policy breaks down into goals for the individual sectors, each of which has its own policies. Although groups call for "national" water policy and many papers and journals write about it, policy usually comes in pieces as legislative bodies deal with issues one at a time. The process ranges from those in democratic countries which may lack coherent strategic plans to those in authoritarian countries where five-year plans may drive policy.

Setting water policy indicates what government intends to do on the basis of values and public choices. Strategy translates policy into action by determining the actions necessary to achieve its goals. Governments use different political systems to determine their priorities, strategic directions, and programs.

Water policy comprises a comprehensive set of policies across the sectors of water supply, wastewater, irrigation and drainage, flood control, instream flow management, groundwater management, and areawide coordination. It breaks down into goals for these individual sectors, each of which has its own policies.

Although many papers and journals report about water policy and groups may call for national water policy, there is usually no unified national water policy, and that is the case in the U.S., where policy comes in pieces as Congress and other legislative bodies deal with issues one at a time. At the national level, U.S. water policy comes mostly through legislation, such as the Safe Drinking Water Act, the Clean Water Act, and environmental laws. By the same token, the U.S. has no strategic plan to implement its water policy. Like the policy itself, strategy becomes apparent as the political process works out, one issue at a time.

The U.S. Clean Water Act is a good example of water policy. It sets a national goal for zero discharge of pollutants into waterways. Sometimes an executive order will give policy direction to government agencies, such as an order to use restrictive measures to protect wetlands. Policy coordination also occurs through non-governmental institutions such as the Colorado Water Congress, which is a membership association involving utilities, agencies, businesses, and individuals and is very influential in the state legislature.

Zero discharge of pollutants into waterways. A goal of the U.S. Clean Water Act.

You can also imagine countries that announce policy and strategy, perhaps in the form of five-year plans. These have the advantage of efficiency but they lack the checks and balances inherent in the U.S. and similar political systems, and it is always problematic as to whether the pronouncements lead to successful action.

Although water policy in the U.S. is distributed across water sectors and levels of government, interest groups continue to call for new policies. For example, a group called the Water Policy Network (2009) wrote this: "A comprehensive integrated national water policy is needed to develop ways to increase water supplies and improve the availability, reliability, and quality of freshwater resources to meet critical municipal, industrial, agricultural, energy, security and environmental needs. The policy needs to coordinate the activity of federal agencies with the States, Tribes, and local public and private entities responsible for water management."

Usually, such calls for a national water policy are aimed at solving some large-scale, intractable water problem of special interest to the group calling for the policy shift. In any case, a single, integrated national water policy seems like too lofty a goal to be obtainable, and policy will continue to be fragmented.

Policy integration in water governance

It is a challenge to integrate water policies across the levels of government to achieve balanced goals at national, regional, or local levels. Often, the approach used is the establishment of a national water council, agency, and river basin organizations. This was attempted in the U.S. through the 1965 Water Resources Planning Act, and more recently was implemented in Brazil through legislation to establish the National Water Agency (ANA) and river basin committees.

In Brazil, water is plentiful but unevenly distributed and the main issues are to balance water uses and to manage the problems of large cities, such as Sao Paulo and Rio de Janeiro. Through its new institutional arrangement, Brazil has made progress toward IWRM with non-partisan professional work and democratic input. A case study presented by the Global Water Partnership (2010) illustrates the need for unambiguous water laws, for strong and well-funded executive agencies, for charging systems that recognize water as a public good, to involve users and the public in decisions, and for basin-wide planning, especially when decisions taken upstream affect water in downstream areas.

The experiment with the US Water Resources Planning Act did not endure as the programs were mostly eliminated during the 1980s Reagan Administration. The reasons are discussed in more detail in Chapter 16.

Empowerment

Empowerment is the function of governance that enables and assists individuals and organizations to provide water-related services and regulatory controls. It is needed mostly for functions that fall outside the narrow missions of utilities and for organizations that lack the capacity to meet their requirements.

The empowerment function of governance begins with how services and their controls are established in the first place. Basic services such as drinking water are usually present in well-established areas, and the issue is to make them work better. However, in developing areas governance deals with establishment of new services and, unfortunately, many people in poverty still lack access to reliable and safe water and sanitation.

The empowerment of a water services organization might stem from authority granted in enabling legislation. For example, in the U.S. local governments derive their authority from state legislation that authorizes what they can do, and they use this authority to establish water services. In turn, the city administrative codes will have sections that authorize the establishment of utilities.

At the national level, a special bill may serve as the enabling legislation for a water agency's program. For example, the US Environmental Protection Agency will have enabling legislation to authorize its actions. Congressional authorization for a project by a government agency will authorize and set purposes of a project. The project only becomes a reality when funds are appropriated for it.

The assistance that water management organizations may receive can be financial, technical or institutional. Financial assistance can be in the form of grants, loans, or guarantees. Financing for a local water project is provided when a governing board approves rates that will finance the project. Technical assistance can be in forms such as training, research, data management, equipment loans, and other forms. It is usually provided by a government agency or an association to small water utilities. Helpers and enablers are found across categories from government agencies to private companies to the non-profit and education sectors. Institutional assistance might provide consulting or special assistance to establish or nurture programs.

The empowerment function of governance includes incentives to encourage desirable behavior, such as meeting needs of low income people and improving aquatic habitat.

Empowerment can go beyond control to provide incentives for positive actions toward sustainable water management.

When you consider varying political philosophies about the roles of government, the empowerment functions of water governance can be controversial. For example, some people believe that government has no role in helping programs such as water supply along, that they should be enabled entirely by the private sector. On the other hand, when communities and individuals are powerless to solve such basic problems, the need for empowerment seems more obvious.

Controlling

The governance process of control seeks to balance water uses to assure equity and to limit actions that might be harmful. Its main form is regulatory control but planning and coordination also serve as controls. In fact, empowerment can be a kind of control as well.

Balancing water uses through planning and coordination is required to deal with our shared dependence on water for many uses, as is explained in detail in Chapters 8 and 16. Because in the hydrologic cycle the same water flows in a continuous way past many users, decisions about its allocation generates many conflicts. To explain these conflicts, the American author Mark Twain wrote: "Whisky is for drinking and water is for fighting."

To see the nuances between the levels of control, you can compare them with police work, with which everyone is familiar. A traffic ticket is a form of regulatory control, but driver education might be considered as empowerment, even if it is also a form of control. Another form of control through empowerment and coordination would be the case where bicycle advocates work as volunteers to coordinate education of drivers and riders for mutual benefit.

As you would expect, there are many categories of water regulation. Usually, they are supported by the authority of statutory law. Examples from the U.S. are given in Table 3.2.

Examples of planning and coordination also show how controls are applied in water management:

- Joint work on a river basin plan
- Inter-city negotiations to share water during emergencies
- Multi-stakeholder work in a watershed to develop best management practices
- Urban planning meetings to decide where to locate stormwater facilities
- Multi-stakeholder meetings to develop a water release schedule for a reservoir
- Shared vision planning to decide on water operations along a river
- Stakeholder meetings to develop a shared river or estuary restoration program
- Meetings among land owners to obtain commitment to keep a watershed clean for drinking water

GOVERNANCE AND MANAGEMENT COMPARED

To illustrate how governance and management should be aligned for the water sectors, we can map water management services, organizations and types of controls required (Figure 3.5). On the left of the diagram are the 10 management services from Chapter 2, in the center are the main management organizations or mechanisms that deliver the services, and at the right are the main required control functions.

Table 3.2 Categories of water regulation

Water withdrawals	The purpose is to allocate supply by making decisions about water for different users or customers, especially during shortages. This can be regulation of water withdrawals through a state agency such as the Colorado State Engineer. The governance needed for water withdrawals is to allocate the supplies needed across the competing uses, including environmental uses. At the smallest scale, such as in a remote watershed, the users might develop a self-governance system that works. However, experience shows that even in cases like this, disputes can arise, such as between upstream and tail-end irrigation water users. These might also conflict with fishers, who seek to maintain instream flows. So, a system for water allocation is needed in almost all cases. Water allocation systems are based on law and management control systems, which are explained in the water supply sector chapter.
Drinking water quality	Regulation of drinking water quality occurs through a federal mandate such as the Lead and Copper Rule
Environmental water quality	Regulation to coordinate upstream and downstream water uses such as when one person's wastewater becomes another's water supply. In this case governance is exercised primarily through the Clean Water Act, which classifies water for its uses.
Instream flows	Instream flow regulation requires governance with multiple objectives and may only work with effective coordination among authorities.
Codes and standards	Regulation through codes and standards that specify a practice such as cross-connection control to prevent an improper plumbing connection to contaminate water. The control would be under the authority of the Safe Drinking Water Act and plumbing codes.
Interstate flows	Regulation of interstate flows through court decisions, such as the decree for the Delaware River Compact and Basin Commission.
Transbasin diversions	Regulation of trans-basin diversions when local supplies are depleted and people look to other river basins for imported supplies. Governance must consider complex issues that go beyond management of a single stream and may require courts to exercise governance and arbitrate among regions.
Self-regulation of irrigation	Self regulation of irrigation systems may occur to solve the "tail-end" problem of irrigation systems where upstream irrigators take more water, leaving too little for the tail-enders. It may require self-regulation of diversions by users, almost like a homeowner's association, or in the case of a larger system it might require ditch riders or a regional authority.
Flood preparedness	Regulation of flood threats where development by upstream landowners exacerbates flood damages lower in the watershed. Governance may require involvement of local or regional governments or even arbitration through legal processes.
Point source dischargers	Ideally, point source dischargers would clean up their wastewater to the point of "zero pollution," but this is not practical so a system to allocate the waste-assimilation capacity of streams is required. This system is based on the Clean Water Act, which is explained in the chapter on wastewater.
Non-point dischargers	Non-point discharges account for the lion's share of stream pollution, so a system to manage them is required. This is a much more difficult problem than point source discharges.
Hydrologic modifications	Hydrologic modifications that alter the flow of water include dams, channelization, dredging, and dewatering. These require regulation or waterways will continue to degrade.
Water uses	The governance of residential, commercial, industrial, and public water users focuses on the need to limit them to the water they need and/or pay for and to control how they manage their wastewater discharges. This introduces an example of how the water utilities must exercise regulatory authority as part of the governance process.

Management services and processes	Management organization or mechanism	Main control function of governance
Public and industrial supply	Utilities	Regulate use/quality
Wastewater	Utilities	Regulate discharges
Stormwater	Urban service	Regulate quality
Irrigation	District/farm	Regulate use/quality
Farm drainage	District/farm	Regulate quality
Instream flows	Dam operation	Coordinate, require ISFs
Environmental water quality	BMPs	Monitor, set standards
Stream flooding	Flood control/FPM	Emergency preparedness
Groundwater	Model/control	Regulate use
Areawide management	Multipurpose	Multi-purpose

Figure 3.5 Alignment of services, mechanisms, and governance

The information on the diagram is explained in more detail in the following (Table 3.3), which illustrates the main mechanisms and functions of management and governance for each water sector:

Table 3.3 Main mechanisms and functions of management and governance by water sector

Water service	Management mechanisms	Governance functions
Public and industrial supply	Provided by utilities and self-supply.	Ensure access to service. Regulate water allocation and quality. Empower low-capacity utilities.
Wastewater and environmental water quality	Provided by utilities and private providers for wastewater removal from buildings and points of water use and for treatment.	Ensure access to service. Operate permit, monitoring and enforcement programs. Regulate individual units. Regulate instream water quality. Empower low-capacity providers.
Stormwater and flooding	Provided by organizations for stormwater and flood control from urban areas and sites.	Ensure service provision. Regulate flood conveyance, hazard zones and nonpoint sources. Ensure emergency preparedness.
Irrigation and farm drainage	Raw water and drainage systems provided by organizations and individuals in irrigated and rain-fed zones.	Regulate water allocation and drainage water quality. Self-govern irrigation organizations. Empower irrigators.
Instream flow control	Coordinate flows for hydropower, navigation, recreation, fish and wildlife and conveyance water.	Establish instream flow requirements and control diversions and discharges.
Groundwater management	Coordinate and manage groundwater development and uses among stakeholders.	Regulate groundwater withdrawals and protection.
Areawide and multi-purpose management	Operate multi-purpose and areawide agencies for coordination and/or river basin planning.	Empower and coordinate actions. Regular sector and other programs, such as interstate compacts.

CONCLUSIONS

The framework for water governance in the chapter is built on a set of institutional arrangements that support the policy, empowerment, and control required by water management. Governance actions differ across the ranges of water services and types of countries, but they remain the cornerstone of a successful control system for water use.

When viewed comprehensively, governance is complex and involves many elements of government and political science. To study it, you must consider the purposes of water use, the condition of institutional arrangements and the levels of government and geopolitical scales. When it is stripped to its essentials for water services, its focus is on regulation as an important tool, along with policy and empowerment of organizations and individuals to achieve better water management. While governance elements may seem abstract, to be useful they must address the practical issues of providing basic services for water supply and sanitation.

With so many complex issues, it is easy to see that water governance extends beyond the boundaries of the branches of government to include corporate, community and self-governance. It has roles and responsibilities for many people who are responsible to manage water and to provide governance over water services and their impacts.

On the government side, the principle of subsidarity applies to governance, meaning that role should be assigned to the appropriate level of government. Although water management is mostly local, decentralization is not always the answer because the capacity might not be adequate. In any case, policies and programs should be articulated to link things well.

Governance by government agencies is the most familiar scenario, but social responsibility that goes beyond regulation may be fostered more at the level of corporate governance. More broadly, governance draws in professional associations, non-governmental organizations, businesses, and private citizens.

Success in governance amidst the many conflicts and challenges of the real world will require courage and wisdom, as well as the theory. The theory of water governance addresses this reality in its explanation of how adequate institutional arrangements are required. These are intended to deal with real world issues such as incompetence, moral hazard, conflicts-of-interest, and corruption. So, at the end of the day, governance arrangements are a starting point in the long quest for sustainable water management.

Water governance is under scrutiny around the world. To illustrate the range of concerns, the UNDP water governance facility studies the political, social, economic and administrative systems that affect the management of water and water services at different levels. They see how water governance based on catchments addresses equity and efficiency in integrated water and services management for water use between socio-economic and environmental systems. Topics they address include water policies, legislation and institutions; how government, civil society and the private sector interact; ownership, management and administration of water and services; economic and financial topics such as price regulation, subsidies, tax incentives and credits; intersectoral coordination, stakeholder participation and conflict resolution; water rights and permits; control of water quantity and quality; role of women; and bureaucratic obstacles and corruption.

Postscript

During the 1970s interdisciplinary work on water management blossomed at Colorado State University. It was an exciting time, and we were fortunate to sit at the feet of people who had "been there" and participated in water management issues around the world. While most of them were engineers or scientists, the political scientists, economists and sociologists also made landmark contributions.

From some of them I came to hear about the institutional arrangements for water management, which was a new concept. Having been through a lot of military training and participated in business, I saw the world as black-and-white, and the experts seemed to see a kaleidoscope of colors that I was missing. I came to wonder, what do they know that I don't?

Although more than 30 years have passed, I really can't answer that question yet because understanding institutional issues is a lifelong process, but I can point to some of the lessons. The people who represented these lessons the best to me were: Henry Caulfield and Norman Wengert, both professors of political science; Evan Vlachos, a professor of sociology; and Charles Howe and Gilbert White, both associated with the Institute of Behavioral Sciences at the University of Colorado. Wengert (1972) wrote widely about institutions, Caulfield had been the first director of the U.S. Water Resources Council and could tell many real world stories of federal water actions, Vlachos has keen insight into organizations, systems, and governance, and Howe and White have made substantial contributions to our understanding of water economics (mostly Howe) and integration in water management (mostly White).

Nobel prize winners in economics also inspire us to think about water institutions. For example, the 2009 winner, Elinor Ostrom, was recognized for demonstrating how property can be successfully managed by local commons without any regulation by central authorities or privatization. Ronald Coase, the 1991 Laureate, studied utilities and wrote a much-cited article on "The problem of social cost," which explained how transaction costs affect behavior and how external costs can be internalized.

So, the study of institutional arrangements involves heavy lifting and it is fair to ask, "How can you apply it?" I think you apply it to practical problems that occur a level or two down from the general concepts. For example, in North Carolina I was confronted with an organizational issue in the environmental agency. The previous director had reorganized the programs from being aligned with water and air to an integrated form with names such as planning, operations, and support. While the motivation for the changes was positive, there were many negative effects. Links with constituent groups were broken, managers were in over their heads, and the workforce was unhappy. The solution: reverse the reorganization and return to the previous form. This enabled us to solve problems of managerial capacity and relationships with our constituents. It worked out well. But this was just one of many issues affected by our institutional arrangements. We had to deal with numerous internal and external political influences, respond to criticism from interest groups, and keep the day by day work flowing in spite of bureaucratic constraints. These were challenging and they show why sometimes good ideas do not get implemented quickly.

The take-away for me from these experiences is that you must be a student of your institutional arrangements, in all of their complexity. You have to be patient and exercise the art of the possible. The bulldozer or take the bull by the horns approach will usually fail in complex institutional environments, which is the case for most water management scenarios.

REFERENCES

Barreira A. (2006). Water governance at the European Union. *J. Contemp. Water Res. Educ. UCOWR*, (135), 80–85.

Global Water Partnership (2010). *Toolbox for Integrated Water Resources Management.* http://www.gwptoolbox.org/ (accessed 1 January 2010).

Lafferty W. M. (2004). Governance for sustainable development. In: *Introduction: Form and Function of Governance for Sustainable Development*, W. M. Lafferty (ed.). Edward Elgar, Cheltenham, pp. 1–27.

North D. C. (1990). *Institutions, Institutional Change, and Economic Performance.* Cambridge University Press, Cambridge.

Rouse M. (2007). *Institutional Governance and Regulation of Water Services: The Essential Elements.* IWA Publishing, London.

UN Development Programme (2010). *Water Governance Facility. What Is Water Governance?* http://www.watergovernance.org/whatiswatergovernance (accessed 6 April 2010).

Water Policy Network (2009). *Wanted: A Comprehensive, Integrated Water Policy.* http://www.waterpolicy.net/index.htm (accessed 7 December 2009).

Wengert N. I. (1972). *A State-of-the-Art Review: Final Report: Urban-Metropolitan Institutions for Water Planning, Development and Management: An Analysis of Usages of the Term "Institutions."* Colorado State University, Colorado Water Institute, Fort Collins, Colorado.

World Water Assessment Programme (2006). *Water: A Shared Responsibility.* The United Nations World Water Development Report 2. UNESCO, Paris and Berghahn Books, New York.

Chapter 4

Water allocation and drought response

This is the first of a series of chapters about governance of the subsectors of water management. While general terms such as policy, capacity-building, and regulation apply to all subsectors, each one requires an individual approach. In addition, the approach will vary with the scenario. After all, the goal of governance is to do what is required to bring about successful water management, not to simply implement government controls for their own sakes.

The kingpin issue of water use is its allocation in the first place. This is the water management function that generates the most conflict and is the subject of most "water wars." Thus, the control system for water allocation is of paramount importance in water governance. At the end of the day it is about sharing water, but the incentive systems to use all you can are powerful and work against equitable allocation. Therefore, the system of governance for water allocation must be like a judge, an accountant, a sheriff and a peacemaker all at once.

Governance for water allocation must be judge, accountant, sheriff and peacemaker.

WATER ALLOCATION: THE GATEWAY TO WATER USES

As a key part of the water sector, water allocation has the most influence on balancing water diversions from natural systems and is fundamental to IWRM. While it is mostly a regulatory function, it is presented as a subsector of water management because it involves distinct operating agencies and, although it exercises control, it requires governance itself.

In a number of countries, this function is handled by national departments of water resources. In the U.S., it is administered mostly by departments of water resources, state engineer offices or state water control boards.

In some other countries, it is handled by regional organizations under delegated authority, as in Colombia, where it is managed by regional autonomous corporations under the authority of national legislation.

The chapter explains how water is allocated from surface water and groundwater sources and how withdrawals are regulated. It explains how water allocation is handled during both normal and drought conditions, which is an important distinction because extreme natural or human-caused drought can create emergency conditions.

Governance of water allocation is required in countries of all income levels because failure to allocate water equitably can lead to hardship, conflict and deprivation. After water is allocated, governance of its use is the key to introducing water efficiency and to overcoming perverse incentives for using water wisely. Governance for drought water management provides a mechanism to respond to droughts of varying intensity.

The chapter addresses governance issues of water allocation and use, but it does not deal with those of water supply utility performance and the safety of drinking water. Those issues are addressed in the next chapter.

THE WATER ALLOCATION SECTOR

Governance of water allocation and use deals with two distinct but inter-related objectives:

1) To allocate water taken from streams and aquifers equitably; and
2) To make sure that water diverters use water responsibly and do not waste it.

These objectives apply during normal and drought conditions. The case of drought is especially important because, in many places, streams are not stressed during high flow periods but they may become so during drought, when equitable and responsive allocation becomes especially important.

Water allocation is most critical during low flow periods.

Figure 4.1 shows the water management scenarios covered by this category of governance. Five diversions are shown, from top to bottom: a well to irrigate a farm; a surface diversion to a city; a surface diversion to a farm; a well for a self-supplied industry; and another city diversion. The water uses correspond to the diversions and are for municipal, industrial, and agricultural purposes. Water entitlements from the reservoir shown might also be governed under the principles discussed here and, if they are, they might involve a combination of water storage and diversion rights (see Chapter 14).

The governance of water allocation and water use involves different types of organizations who must cooperate. Water allocation is normally based on state or national laws, and is controlled by regulatory authorities at the state, provincial, and regional levels. Water use authorizations are usually controlled by local authorities who distribute the water, such as utilities or water supply companies. These local authorities are usually responsible also for water distribution and safety, and the next chapter outlines their main responsibilities. Their involvement in water use governance may be in the form of their water efficiency, conservation and drought-response programs.

To visualize the separate facets of water allocation and use, think of the water master who controls the allocation of water to distribution companies, who provide the water to end users. Figure 4.2 illustrates this division of functions and how the water allocation sector relates to the water supply sector.

Allocation from natural water systems

The water allocation sector is responsible for water withdrawals from renewable surface sources, from aquifers, and from stocks in reservoirs. To get an overall view of the water demands and water withdrawn, data at the basin and higher levels are needed. To know about the supplies of natural water available, hydrologic data are required. Therefore, both hydrologic data and water use data are required to manage and govern this sector.

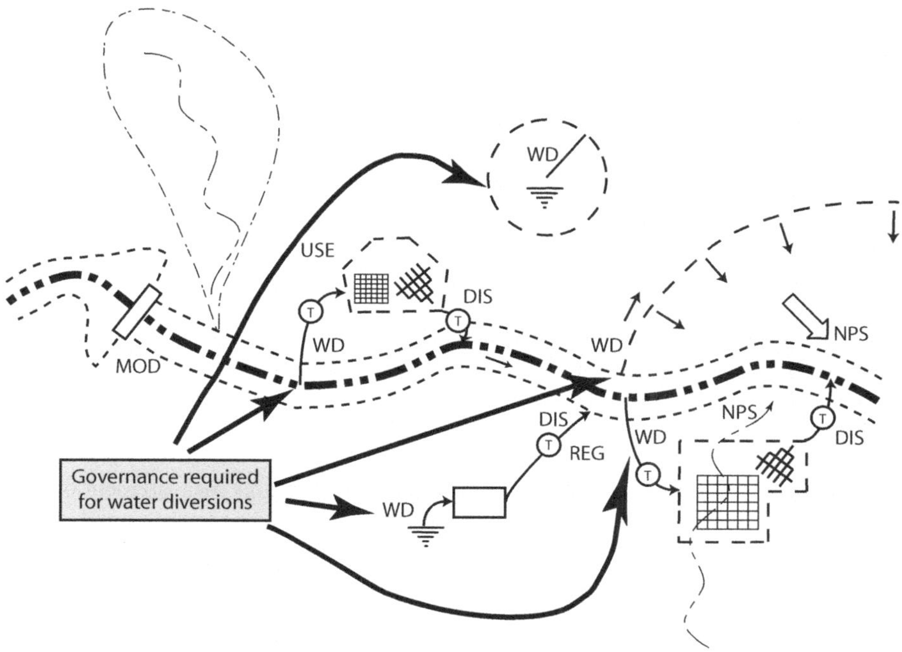

Figure 4.1 Water diversions that require governance controls

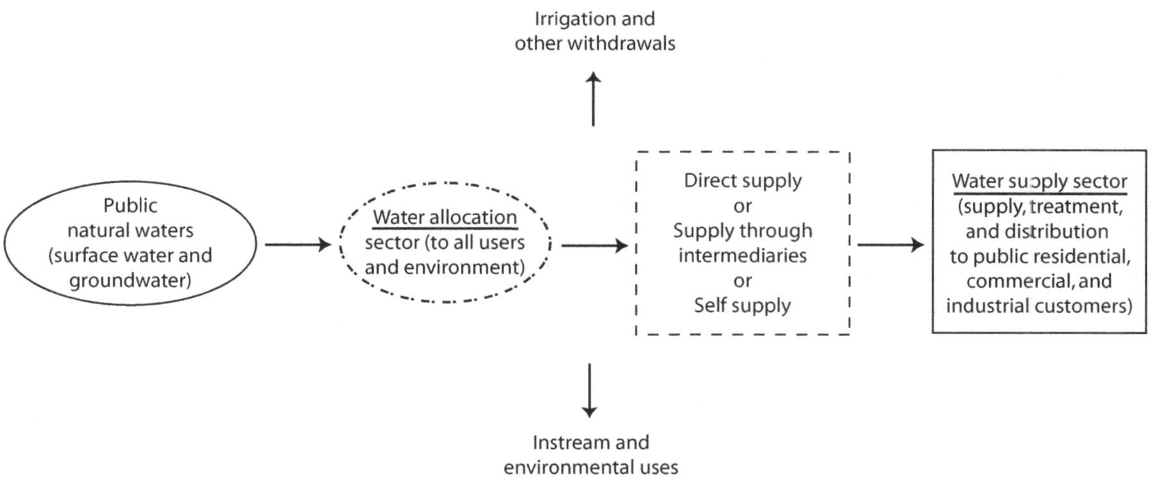

Figure 4.2 Water allocation and public supply sectors

For example, in the U.S., both hydrologic and water use data are reported by the U.S. Geological Survey (Barber, 2009). Surveys of water withdrawals are taken every five years, and the totals have changed little for more than two decades. On the basis of withdrawals in the U.S., public supply is about 11%, irrigation about 31%, self-supplied industrial water 4%, and thermoelectric water about 49%. Other smaller uses are for homes on wells, livestock and aquaculture, and mining purposes.

As explained above, sector membership includes the water allocation authorities and the water distribution organizations who may control use of water. Sometimes wholesale water supplies are provided by intermediaries, such as special districts or government agencies. These are discussed with the areawide water sector (Chapter 10), including government water agencies such as the U.S. Bureau of Reclamation and U.S. Army Corps of Engineers. Sometimes, a water utility will provide both wholesale and retail water services, and these are discussed in Chapter 5. If water is provided from one utility to another, these are called "consecutive systems."

Related to state water quantity regulators for surface and groundwater are policy makers for development and allocation of water (discussed in Chapter 16), state water planners and districts for regional water development and management (see Chapter 10).

As shown in Figure 4.2, water allocation functions begin with the native environmental water supply. Before any allocation can be made, hydrologic data must be examined to assess the supply to determine the reliability of water yields. Then, exogenous demands are identified and assessed to determine their qualifications and priorities. The allocation takes into account the priorities, rules, and practical matters of managing the supply for all uses.

After the supply is allocated, the system must be monitored. This requires measurement of the native supply, the withdrawals, returns, and losses. These may lead to administrative actions to reassess demands and in some cases to reallocate the available water.

In the event that emergency action is required, the water allocators may impose short term rules to reallocate water supplies according to a drought response plan, which is a kind of emergency response plan.

Governance of water allocation requires drought response plans.

The regulators discussed in this chapter are those who control the allocation of environmental water. The regulators who control the quality and safety of water supply services are discussed in Chapter 5. While these functions are closely related, the regulators are often in separate organizations if not in separate sections of the same organizations. In other words, integration of water quantity and quality is more difficult than meets the eye.

Water allocation and water supply organizations serve different customers. Water utilities serve individual customers, whereas water allocators serve the public at large and make decisions on the basis of the public interest. The focus of the water allocation sector is on raw water quantity and assuring a balanced allocation of environmental water to all qualified users. The role of the water supply sector is to develop water supplies for distribution to their utility customers.

The water allocation sector manages publicly-owned natural waters. Water supply utilities manage water that they are licensed to distribute to individual customers. They also are responsible for the public health aspects of their distributed water, whereas the water allocation sector focuses mainly on water quantity. Sometimes water allocation regulators are lumped with environmental water quality regulators, but usually their functions are separated, at least in the U.S. The environmental water quality regulators are discussed in Chapter 6 with the wastewater sector.

Regulators of water allocation sometimes have responsibility to manage environmental water for habitat. In that sense, they are closely linked to the instream flow sector, which is explained in Chapter 8. Water supply entities usually have only an indirect responsibility for water-based habitat and may consider rules about environmental water to be obstacles to gaining access to water supplies.

GOVERNANCE ISSUES

The water allocation process involves the governance tasks shown in Figure 4.3. Explanations of them are given in Table 4.1.

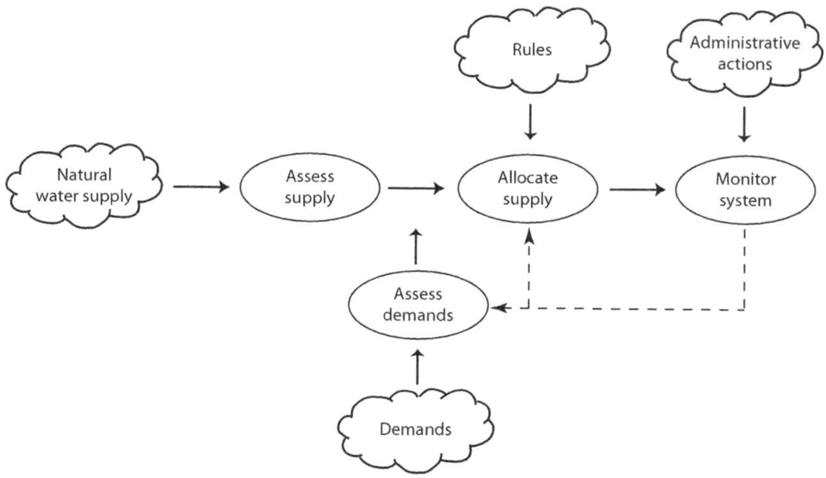

Figure 4.3 Water allocation functions and roles

Table 4.1 Governance tasks in water allocation

Function	Governance issues
Assessment of supply	Regulators need valid hydrologic studies to establish the reliability of supply.
Assessment of demands	In a permit system, demands are evaluated in terms of reasonableness. In an appropriation system, demands depend on the ability to apply water to beneficial use (see Chapter 14).
Rule making	An enabling statute is required to establish authority to make rules about water allocation. The rules contain elements of the management system.
Allocation decisions	Allocation decisions are made according to the system (usually a permit or appropriation system, see Chapter 14).
Administrative actions	Operating the regulatory system requires officials such as ditch riders, basin authorities, and a central authority.
Monitoring	Gaging stations and an information system are required to collect and disseminate data on water availability.
Emergency response	A drought response plan is required to handle emergencies. If a disaster such as contamination occurs, rendering supplies unusable, it can be handled by the same plan.

Water allocation law is required to set policy for water allocation because without clear and strong rules, conflicts may get out of hand. Allocation of raw water is through water rights or permit systems. In the U.S., these are instruments of state governments, and their frameworks differ from state-to-state. The U.S. has two basic systems of water allocation law, one based on the riparian doctrine (which is administered through permits) and the other on ownership of water rights. These are discussed in Chapter 14.

It is a challenge to have fair systems for water allocation for countries at different levels of development and different stages of the Rule of Law. For the lowest-income countries, a top-down system of water administration may be needed until they are ready for more advanced systems. Water allocation may have to take into account purposes such as having minimum quantities of irrigation water for subsistence farming. After such basic purposes are satisfied, a society will turn its attention more toward meeting more discretionary water needs.

Whether the water allocation system is based on permits or ownership of rights, water users require security of their rights to use water; otherwise, they will not have the incentives to invest in long-term infrastructure facilities.

Governance of water allocation provides security for water right owners and permit holders.

In the water allocation sector regulators must be empowered to control, facilitate and monitor the supplies of water to users. This assumes that the users have the capacity to divert and manage their water, and the water allocation regulators are able to control diversions and returns. If any of these capabilities are weak, the empowerment role of governance becomes more important.

Being a regulator in this sector can be a tough job. It is like a combination of a judge and a sheriff. The regulatory function decides on allocations (judge's role) and administers the law (sheriff's role). It requires scientific knowledge, judicial wisdom, and the backbone to enforce the law.

For water users, empowerment may mean to equip them with education about water efficiency and perhaps to assist in implementation of systems for metering and water monitoring. In all cases, the regulatory agencies and water users should be educated about the need to work together to maximize benefits to all.

Regulatory controls, along with planning and coordination of water withdrawals and returns, are required to implement the water laws. Chapter 14 explains the principles of water allocation laws and rules.

The environment is a different kind of water user, and the regulator's job is to prevent it from being "robbed" of its water, so to speak.

Water allocation governance looks after the environment as well as human uses.

WATER ALLOCATION SCENARIOS

This section provides a few examples of governance situations that arise in water allocation and use.

In the U.S. it is common for eastern states to follow the riparian system of water allocation law, but this has limited application because users cannot in a practical sense receive the flow past their property "undiminished in quantity or quality." This issue is explained further in Chapter 14, but to respond to the practical needs, states normally develop administrative permit systems to allocate the available waters. The procedures for this are also explained in Chapter 14.

It is much less common around the world to allocate water in an appropriation system, which requires an extensive system of laws, regulations, procedures, and staff. Water administration in the State of Colorado is an example of how such a system can work. The details of the system are explained in Chapter 14 and, although in theory an appropriation system may work well, the day-to-day functioning of the system may generate conflicts. As long as water is relatively abundant, as in a wet period, conflicts are at a minimum, but when supplies run low, the competition and conflicts over water will increase. For example, in Northern Colorado a system of surface water and wells supplied municipal and irrigation users for decades, and a system of rules was implemented to prevent well-pumpers from depleting surface supplies. However, it failed to balance supplies and many well-pumpers were forced to curtail their water use, creating hardship in the agricultural community. This illustrates the "winner take all" aspect of the appropriation system, as compared to the more balanced approach of permit systems. The other side of the coin is that some consider a permit system to be based on "czar rule," and they do not like the idea that an appointed official can determine water rights and security.

As discussed in Chapter 13, transboundary water issues occur when water moves across political jurisdictions beyond the reach of a single authority that is empowered to balance the competing claims on water resources. So, they can involve water allocation issues such as surface or ground water allocation across borders or interbasin transfers. In these cases the lack of a unitary authority to allocate the water means that some agreement or compact will be required to resolve disputes and achieve allocation. The mechanics of these are discussed further in Chapter 13.

In some cases, the same water authority that has authority to allocate water may have responsibility to develop water sources. An example would be a state or national agency that has all responsibility for management of water, such as a Ministry of Water Resources. The apparent conflict of interest can be resolved at least partially by the realization that the agency is not a water user but is in fact a wholesale water provider as well as a regulator. As long as the agency has a monopoly over water development the dual role does not give it any advantage over competitors. With any such arrangement, the problems of monopoly management should be confronted and mitigated.

Irrigation water must be allocated to users from big streams and, once the water is in conveyance ditches, it must be allocated to end users. Chapter 7 explains some of the problems that arise, such as the "tail end" problem where if upper irrigators take too much, then there is no water for the lower irrigators. This problem requires an allocation scheme with management and enforcement. It can range all the way from large and formal systems to small self-governance situations.

Water use control is required to achieve water use efficiency and confront ever-increasing demands in the face of supply limitations. The allocation process is itself a form of water use control but once water is allocated, there is a need to impose additional controls on water use so that users have incentives to use supplies efficiently. This problem gets back to the fundamental issue of whether water supply is a right or an obligation. If it is a right, then users will try to get all they can from their supplies without voluntarily cooperating with the need to release water to others. If it is an obligation, then users would be expected to take only what they actually need and release any excesses to others. Experience shows that users respond more to the concept of water as a right and must either be required or incentivized to use water efficiently and release their excess water. These concepts underlie the continuing search for the best approaches to conservation and water use efficiency.

DROUGHT RESPONSE

Regardless of the legal system to allocate water, special conditions may apply during drought periods when supplies do not satisfy all authorized uses. Normally, a drought response plan is needed to pre-plan how drought emergencies will be met. Depending on the water allocation system, the drought plan may be automatic (as in the case of the appropriation system) or administrative (as in the permit system).

In the appropriation system, the slogan "first in time, first in right" applies, and junior water users lose their access to water, while seniors obtain their full allocations. This system seems unfair, but is a legacy of the property rights concept that underlies this doctrine of water law. A permit system is more flexible, but it requires fair and informed administrative decisions that work well during drought.

While a permit system can accommodate drought hazard, it can become complex to administer and officials may be reluctant to impose restrictions on water use because they cause conflicts. This was the case, for example, in the development of a "capacity-use" approach in North Carolina. The term "capacity-use" refers to limits on use to respond to the water capacity available. This system requires "critical use studies" that might or might not lead to the designation of "capacity stress areas."

It can be difficult for a government to decide on a plan for water restrictions during drought. In North Carolina, for example, the state's Division of Water Resources went to extraordinary lengths to study when streams become under stress and when restrictions should be applied. This was the case when I worked there, and 30 years later the state is still considering water allocation legislation to meet the needs during dry flow periods.

The critical use study and designation of capacity stress areas provide mechanisms to coordinate use of water in an area when there is not enough for all users to take all of their permitted quantities. The program is coordinate basin users to work out shared needs under shortfalls due to drought, overuse, or scheduling problems. Implementation of such a plan will normally specify an administrative trigger mechanism. A problem could be as small as a few water users in an area surrounding a local aquifer or as large as a river basin. An area could include portions of more than one basin, particularly if interbasin transfer is an issue.

CONCLUSIONS

The water allocation sector is important because it has the authority to balance water diversions to meet needs of water users and natural surface water and groundwater systems during normal and drought conditions. The sector is also responsible to make sure the environment gets enough water. It is distinct from the water supply utility sector, which deals with safety and reliability of water supplies.

This sector can be difficult to organize and manage, and it involves different types of organizations who must cooperate. Its authority is normally based on state or national laws, which are based on either a permit or appropriation system of water law. It requires clear rules and transparent administration to keep conflicts under control. Regulators must be empowered to manage, facilitate and monitor the supplies of water to users, and this can be a difficult job that involves judicial and police type work.

A range of governance situations are cited in the chapter, to include: developing a permit system, allocation in the appropriation system, transboundary issues, water allocators as water developers, irrigation water distribution, water use control and conservation, and drought response plans. These are only a few of the many governance scenarios that arise, which require governance responses that are appropriate in the context of each individual situation.

Postscript

The challenge of water allocation came onto my radar screen soon after we arrived in Colorado, which had practiced the "first in time, first in right" system for about 100 years at that point. In fact, the right to appropriate water is embedded in the state's constitution and the dome of the capitol has a series of murals that display our history in and dependence on water. The famous novelist, James Michener, even saw fit to use the Colorado system as one of the centerpieces of his book *Centennial*, which became a television mini-series.

If the West exhibits the truth of Mark Twain's observation ("whisky is for drinking, water is for fighting"), then the East just took longer to start its water wars. Some of them actually started many years ago, such as the Delaware River conflict that ended with a 1961 Supreme Court Amended Decree. During the 1970s the water wars on the Chattahoochee River through Georgia started up (see Chapter 13), and the 1980s drought in the Southeast fanned the flames of water conflicts in several southern states. I learned how these states approach water permits in North Carolina and Alabama and was amazed at how much resistance there is to having the state government involved in issuing water use permits. The compromise position seems to be that the government can be involved during drought but is advised to use a very light touch during normal times.

Of all the lessons I've learned about water allocation, the one that stands out the most is how tightly people hold on to the right to use water. It's to be expected in the West, where the right to use water carries value that reaches higher than the value of land that the water is used on. For example, in Colorado the water rights to irrigate an acre of farmland might be worth three times as much as the land. That is not always the case but it often is. People in more humid zones also hold tightly to the right to use water, even when they are not able to sell the right as they can in the West.

In Colorado, the water rights system is in many ways a have and have not situation. We have seen that in a conflict between senior and junior water right owners where the seniors filed a suit that resulted in well-pumpers being cut off from their supplies, causing great hardship among many small farmers. This conflict has proved very tough to ameliorate.

Viewing the pluses and minuses of systems of water allocation has shown me that no system is perfect. The appropriation doctrine creates the have and have not situation and seems not fair at times. No one likes a government czar calling all of the shots in a permit system either. What is needed is a system of cooperation where water users can work out issues themselves. That is what Elinor Ostrom was looking for in her research leading to the Nobel Prize in economics (see Chapter 3), but how it can apply in conflict-filled water allocation scenarios remains to be demonstrated in most cases.

REFERENCE

Barber N. L. (2009). *Summary of Estimated Water Use in the United States in 2005*. U.S. Geological Survey Fact Sheet 2009–3098, Washington, DC, 2 p.

Chapter 5

Water supply services and regulation

The water supply sector meets the most basic needs of people for safe drinking water, and its utilities provide water for multiple other needs of society. This is the most important subsector of the water industry for looking after public health and the daily needs of homes, businesses and public life. It is under much pressure as supplies dwindle, contaminants increase, and demands grow.

If it did not face enough pressures, the water supply sector is increasingly being required to consider water needs for the environment as well as for cities and industries. After environmental water is allocated, the water supply sector processes and provides water to homes, businesses, factories, and other public and private facilities. In addition to potable water, the sector's organizations also provide raw and industrial-grade water to some customers, and are increasingly providing reclaimed water for multiple purposes.

Scenarios in the water supply sector range from the smallest villages to the largest and most sophisticated utilities. Large international businesses are active in the sector, many of them being spinoffs from public companies that went private. The dominant organizational form is the small water utility serving towns and built areas with limited populations. These small utilities are required to provide water of the highest quality to safeguard public health and often they lack capacity needed to do their jobs. The international goal to expand safe drinking water service faces this dilemma to improve public health around the world.

WATER SUPPLY: THE MOST BASIC WATER SERVICE

The foundation of the water supply sector is public health protection and its governance must be of high priority at all levels of government because safe and reliable water service is needed to support life at many levels. Water managers can take pride that improvement of water quality over the past century has been responsible for large declines in water-borne diseases such as cholera and dysentery.

Safe water supply is critical to public health.

The major players in this sector are urban water utilities, which are found in cities across the world. They range from water departments in small towns to giant public utilities, such as the Los Angeles Department of Water and Power. Many of them are public companies owned by shareholders, such as the water companies of the U.K., which were created by privatizing regional water authorities. The sector's members range from these large and sophisticated business-oriented utilities to small entities that serve only a handful of people and which may not even employ a full time worker. They may be government-owned or privately-owned. The sector also involves many self-supplied end users.

In higher-income countries, water supply services are expected to work effectively and reliably, and governance can focus on improving regulations, studying new risks, assisting lower-capacity providers, and dealing with challenges such as aging infrastructure and climate change. However, in developing nations, many people simply lack access to safe water and the capacity of water providers and individuals to meet the basic needs falls short of the level needed.

Regardless of the country, governance of the water supply sector has the highest profile among the water sectors because of its link with public health and the standard of living. In addition to its use for drinking water, a reliable and safe domestic supply of water is required for household hygiene to social uses such as recreational swimming.

This chapter addresses governance of the water supply sector across the spectrum of countries. It covers all aspects of water supply service and its regulatory controls. Governance of water quality and safety is usually sharply focused, and the chapter explains models of regulatory controls. Other areas of governance, such as for service quality and business practices, are not as well defined. Levels of service are controlled through mechanisms such as fire insurance practices, codes and standards, engineering practice, and related institutions. Policy for business affairs differs among public and private sector organizations. Customer relations management is controlled more by the water supply industry's benchmarking and quality control practices than by any central authority.

CHARACTERIZATION OF THE SECTOR

The water supply sector is sometimes called the municipal and industrial (M&I) sector because its urban and business customers use piped water that is delivered by utilities. However, substantial quantities of self-supplied water are also provided to a diverse group of end users who do not depend on utilities.

In terms of volume, the main commodity distributed by the sector is treated water for potable, commercial, and industrial uses, but the sector also involves a large point-of-use treatment and distribution sector, as well as the premise plumbing systems that comprise the "last mile" to end users. Bottled water can also be considered a part of the sector, but it is generally regulated outside the sphere of public water supply.

It might be useful to categorize water supply subsectors as utilities, large self-supplied users, small self-supplied users, premise plumbing, and bottled water. Each of these has different attributes in terms of governance, equipment used, and organization. The regulators comprise a separate group in terms of their governance.

The sector's elements can be viewed by their position on the water delivery chain (Figure 5.1). At the points of diversion from water sources, members of the sector operate many types of systems to access, divert, and transport raw water using facilities that range greatly in size and capacity. Source water facilities can range from massive dams and reservoirs such as taking water from Lake Mead or the Great Lakes down to small wells and pumps that might serve only small systems. Then, water is treated by facilities ranging from sprawling complexes to small package plants serving only a few customers. A large treatment plant might produce 100 mgd and occupy many acres of land, while at the other end of the scale, a small shed may house a well and disinfection unit to supply a rural community.

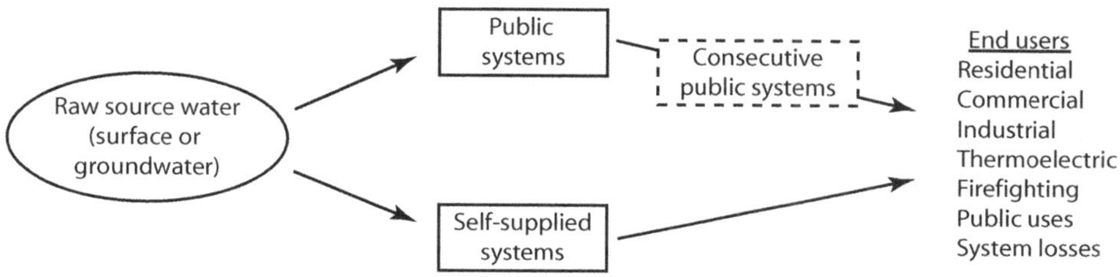

Figure 5.1 The water supply and use chain

To distribute water, U.S. municipal distribution systems have more than a million miles of many types and sizes of pipe and millions of appurtenances such as valves and meters. Water may also be distributed through small, decentralized pipe systems, even from small sources such as individual wells connected to a few customers.

Public water supply requires vast infrastructure systems.

The final elements of the systems serve end users with premise plumbing, household fixtures and appliances. These are normally the customer's responsibilities, rather than of the utility. The premise plumbing systems of end users are regulated in different ways than utilities. These differences will be explained later in the chapter.

On Figure 5.1 you see public water systems, which in the U.S. supply most of the water, and self supplied systems, which are used by many residential and industrial facilities. The public systems may distribute the water through the consecutive systems, which normally re-sell water from primary utilities to their customers. For example, a large urban utility like the Chicago Department of Water will sell wholesale water to a number of smaller, mostly suburban utilities.

Wholesale water may also be provided by regional or state entities, which are discussed in Chapter 10. A large self-supplier, such as a process industry using more than 0.1 mgd, may act as its own utility and require permission to withdraw water. If a self-supplied user such as a rural residence only requires small quantities, it might be exempt from regulation.

The uses of water shown on Figure 5.1 are general categories that provide us with a classification system for water uses based on types of end users, including residential, commercial, industrial, firefighting, other public uses such as street cleaning, as well as system losses, which can be substantial. Use of water to cool electric power plants is shown in a separate category, although it can also be considered as industrial use. Some irrigation water supply is also provided by the sector, but irrigation is discussed as a separate sector (Chapter 7).

WATER SUPPLY ORGANIZATIONS

The service providers in the water supply sector are mainly utilities that provide water on a cost-reimbursable basis. The USEPA (CWSS, 2010) maintains data on utilities and has named them community water systems. For regulatory purposes, EPA defines them as public drinking water systems which serve at least 25 people or 15 service connections for at least 60 days per year. EPA does not regulate wells that supply water to fewer than 25 people.

The latest count of these community water systems shows about 53,000 in the U.S. These range from large big-city utilities to small systems that serve just a few customers. In addition, many non-community water systems, such as campgrounds, highway rest stops, and other transient and non-transient uses are covered.

The U.S. has about 53,000 Community Water Systems serving an average population of about 6,000 each.

Figure 5.2 illustrates the approximate numbers of water systems in the U.S. On the figure, you can see the community and non-community systems and who they serve, and you can see the individual systems that serve people not connected to networks or organized systems.

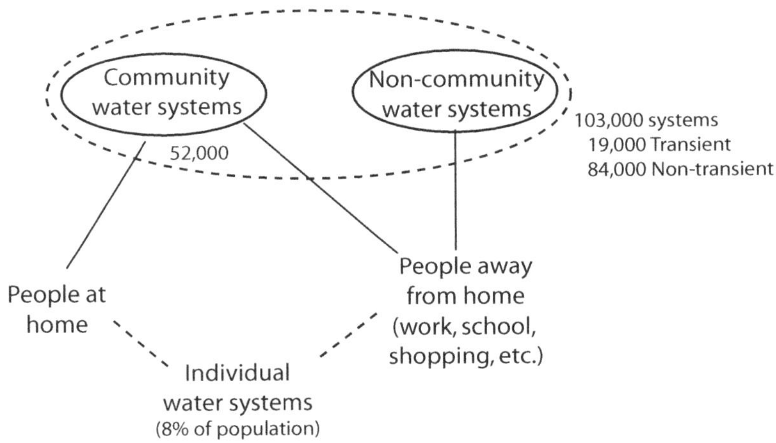

Figure 5.2 Where people get water supply in U.S.

Of the some 53,000 US water supply utilities, 56% serve fewer than 500 people and 92% serve fewer than 10,000 people. The smallest ones, serving 25–500 people, normally have one to two employees, use groundwater or purchased water; employ no operator, engineer, or laboratory, and have active involvement of board members. Even larger utilities in the category serving 3,301–10,000 would only have 3 to 8 employees and a groundwater or purchased water source, perhaps a small treatment facility with an operator, but they would usually employ usually no engineers or laboratory personnel. They would also experience active involvement of board members or function as a city department.

Governance of small water systems is probably the most important issue of public water supply regulation. For the U.S., the data show that over 50% of all utilities have only one or two employees, about 85% of utilities have three or fewer employees, the very small or small utilities only serve some 11% of the population, and are located mostly on urban fringes, in small towns, or in rural areas. Many of the systems serve areas with lower than average income levels and do not expect high water rates. The role of governing boards for these small utilities is more “hands-on” than in most larger utilities (Grigg and Zenzen, 2009).

Figure 5.3 shows characteristics of the large and small systems. Although the block diagrams look similar, the scale of operations of these systems is distinctly different.

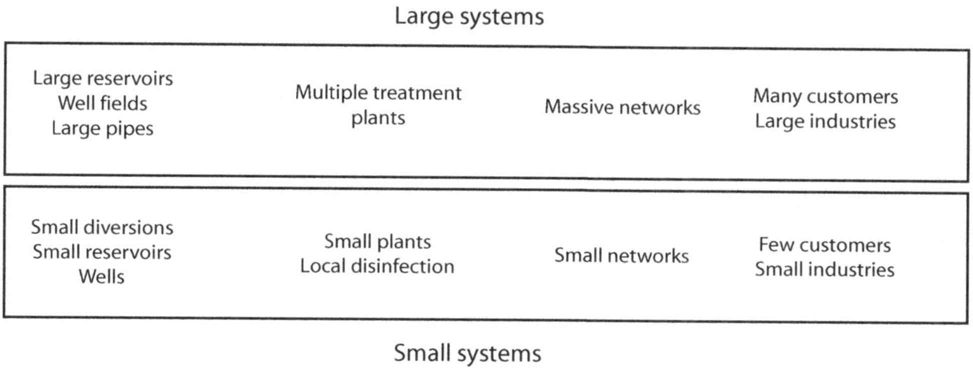

Figure 5.3 Comparisons of large and small water supply systems

While water supply regulators may control water rates as well as water quality, the main focus is on public health. Regulatory issues such as infrastructure standards and performance levels are part of the overall concern about public health, even if they are controlled by different rules, such as city codes, as opposed to health department rules.

The main regulators are state environmental protection agencies (state EPAs), which are the front line troops to enforce the SDWA in the states. Depending on their institutional history, these may regulate both water supply and wastewater. Drinking water regulation evolved along the path of public health policy and was in health departments down to the county level. Wastewater regulation came along later and after the Clean Water Act was passed, it was usually located initially in state environmental departments, which were mostly separate from the health departments. Over the years it became apparent that water and supply and wastewater regulation involved many similar functions and issues, and these were combined in many states.

State public utility commissions (PUCs) regulate the business affairs of privately-owned utilities. The number of privately-owned utilities varies with the region, but on an overall basis most water is supplied in the U.S. by government-owned utilities.

WATER SUPPLY GOVERNANCE

Policy

Policy for the water supply sector should address issues of utilities, large and small self-supplied users, premise plumbing, and bottled water. In actuality, these are quite different and require policy for safe drinking water, industrial-grade water, wells, the business of water conditioning, plumbing rules, and regulation of bottled water.

At the national level in the U.S., the main challenge is the complex problem of developing and sustaining a massive networked system that taps water from many sources, processes it, and delivers it safely and reliably to hundreds of millions of customers. Fortunately, the water supply sector is not a centralized business that requires coordinated decisions and controls from the national level of the U.S. On the contrary, it is a decentralized, mostly-local business with many checks and balances, so the national government does not have to worry about every outpost in the sprawling network.

Water supply is mostly a local business subject to national rules.

Oversight of the system is performed at the state level, which is close enough to the local water providers to supervise them on a day-to-day basis. For the most part, national policy is adopted (and sometimes rubber-stamped) by state governments, who are delegated the authority and responsibility to oversee drinking water programs.

So, this sets the alignment of responsibilities for the three levels of government. Keep in mind that the U.S. is much larger than most countries and for a smaller country, it might be appropriate for the national government to assume both the national role and those that U.S. state governments take on.

Experience shows that water safety cannot be left to local providers alone, but requires regulation that in the U.S. has evolved from the earliest days of the Public Health Service to today's complex Safe Drinking Water Act (SDWA), which is based on a multiple barrier approach to protecting water safety. This means that source water, treatment, distribution, and end-uses should be protected to achieve a fail-safe system condition. Therefore, policy must be multi-faceted to address the elements of these barriers.

The history leading to the 1974 enactment of the Safe Drinking Water Act explains how the public health aspects of safe water have evolved and how the legislation is the organizing vehicle for policies that range across source protection, water treatment, and distribution systems, as well as utility and regulatory responsibilities. The Act set forth a comprehensive regulatory program of maximum contaminant levels, reports, enforcement, and other arrangements. Amendments to the Act continue to serve as a dynamic platform for policy reform.

For example, the 1986 Amendments set goals to add new contaminants to the list, especially organics and chlorine disinfection byproducts. It also banned lead pipes and lead solder in new systems, and required rules for monitoring injection wells below water sources and a state wellhead protection program. All surface water systems were required to use disinfection and filtration. The 1996 Amendments created a state revolving loan program and rescinded a requirement to develop standards for 25 new contaminants every 3 years. USEPA was required to consider risk and benefits to implement standards. Community water systems were required to publish consumer confidence reports about their performance in meeting standards. A Source-Water Assessment Program was also added.

Presently, USEPA is studying new regulations, with a special focus on the effects of water distribution systems on water quality.

Other policies of the water supply sector, such as industrial-grade water, wells, the business of water conditioning, plumbing rules, and regulation of bottled water, are much more diffused than safe drinking water policy and will not be discussed in detail here. In the future, the emergence of reclaimed water as an industrial-grade source and regulation of plumbing systems will probably receive more attention.

Empowerment

Empowering the water supply sector means raising the capacity of institutions and the workforce, as well as providing necessary resources. It focuses on the service providers, but regulators must be empowered too.

Establishing water supply services in the first place is an empowerment issue, especially in developing countries or small communities anywhere. In the U.S., water supply enterprises can be established by the public or private sector, and many businesses provide services and equipment as well as sale of bottled water. Some entities, such as factories and rural buildings, can be self-supplied. Given the large number of small water supply utilities, regional planning and cooperation can improve reliability and services.

In the U.S., capacity building is a shared role among government levels and the private sector. USEPA provides help from the national level, but direct contacts with utilities are normally at the state and local levels. For example, the Colorado Department of Public Health and Environment (CDPHE) (2009) has a Safe Drinking Water Capacity Building Unit to provide training, technical assistance, and management support services to public water systems. The unit engages in coaching and assistance, benchmarking, peer mentoring, training, and other forms of technical assistance.

Technical and financial assistance are needed to help utilities and other providers to develop adequate systems and to comply with regulations. In the U.S., financial assistance is through the DWSRF program that is operated by USEPA. In some cases, small water organizations may qualify for state programs for rural and/or small systems.

Technical assistance focuses on small and rural water systems and includes programs such as training through EPA and AWWA and rural water associations. Another form of technical assistance is government- or industry-funded research into health effects and how to make water supply systems work better. USEPA has an extensive research program, and the Water Research Foundation also has a substantial program. Research topics vary over the years, and the Water Research Foundation (2010) has organized its programs into categories to cover management and customer relations, water resources and environmental sustainability, water quality, and infrastructure.

Empowerment can occur through mutual aid where a utility coordinates with other utilities for security, environmental protection and efficiency requires empowerment. Partnerships, regional cooperation, and mutual aid for emergencies add to utility capacity.

Control and regulation

The main governance mechanism for safe water is regulatory control, which in the U.S. comes via the authority of the SDWA. It was explained under the policy discussion above, and its regulatory arm works through standards, reports, and enforcement. The scenarios in the next section deal mostly with regulatory issues and serve to explain further the mechanics of the regulatory approach to assuring safe water.

Most countries have a safe drinking water law and the World Health Organization also maintains drinking water standards.

GOVERNANCE SCENARIOS FOR WATER SUPPLY

Rule making

The SDWA requires a continuing process of rule-making as new issues are discovered and studied. New regulations are the responses to findings about water safety, but imposition of a new rule comes with a cost so the process must be approached carefully. Initially, the focus with the SDWA was on treatment rules, such as identifying contaminants to be regulated. Later, rule-making extended into new contaminants and control programs, such as source water protection.

Although the quality of distributed water is mainly controlled by the treatment process, the focus on regulations on distribution systems is increasing. The starting point in the US was the 1992 Lead and Copper Rule, which was introduced to limit the levels of toxic lead in water supplies. Now, the Total Coliform Rule is evolving to represent a coordinated approach to control of drinking water quality all the way to the customer's tap.

Enforcement

The SDWA requires reports by utilities of the effectiveness of their treatment and water distribution systems. When violations occur, such as an exceedance of a coliform standard, utilities must take certain actions, such as issuing a boil water notice. When utilities do not comply with regulations, enforcement proceedings include notification, investigation, follow-up actions, and in some cases penalties. See Chapter 15 for general information about enforcement procedures.

Source protection

Protection of the quality of source waters occurs through the combined effects of several programs. Access to source waters is explained in Chapter 4 about water allocation, but the quality of these waters requires protection in several forms. The responses in the SDWA are programs for source water assessment and wellhead protection. On the wastewater side, watershed protection takes on several forms, such as control of point and non-point sources to protect water quality (see Chapter 6), and instream flow programs also serve to protect water sources (see Chapter 8).

Utility oversight for operational performance

While health and safety are regulated by SDWA provisions, the arrangement for regulation of service delivery and business performance is through PUCs for privately-owned utilities and self-regulation through the political process for publicly-owned utilities. Many utilities are very small and, while they are required to meet safe water standards, there is not really any effective regulation for business affairs and control is more through empowerment function of governance where the focus is on government assistance to help them meet standards (see the discussion of small water systems in a following section). On an overall basis, oversight is shared among external regulators and the process of corporate governance. Utilities can be regulated for quality of water, pressure and flow capacity, rates, and condition of infrastructure.

Non-utility suppliers

Many industries and other entities have self-supplied water services and are not connected to community water systems. A factory or business may use self-supplied water for its processes, but if it supplies water to more than 25 people, it becomes a public water system to be regulated by USEPA and called a non-community water system. There are about 20,000 of these non-transient systems (like factories or schools) in the U.S. In the same category, there are also about 85,000 transient systems (like campgrounds or gas stations).

Public water uses

Utilities must provide water for public uses, such as fire-fighting, street cleansing, and system maintenance. Normally, no one pays directly for these water services and they are considered as a public obligation. If the utility is government-owned, this can usually be arranged by simple agreements, which may not even be formalized. If the water provider is a different entity than the general purpose government, it might require some oversight through the regulatory process, especially for private water companies. Mechanisms exist to transfer funds within government entities to compensate for such uses and a general purpose government can pay negotiated rates to a private water company for its water uses.

Individual systems

Individual uses are usually homes or other free-standing places where less than 25 people are served. Other than rules on well construction, people drawing water from their own wells are responsible to monitor their own water quality. The number of people supplying their own water in the U.S. is difficult to estimate, but the 2000 Community Water System Survey showed that community water systems served 258.5 million and, as the population that year was about 281 million, it indicates that about 22 million people or about 8% of the population supplied their own water. USEPA does not regulate individual systems, but state health departments regulate construction of wells. Well owners can get some assistance from county health departments as, for example, advice and perhaps water testing from a local health department. The National Ground Water Association (2010) maintains a website to provide information for individual well owners. The problem of individual services is acute in some developing countries, where the people lack the capacity to construct and maintain their own systems.

Premise plumbing systems

Although premise plumbing systems are generally the responsibility of building owners rather than utility providers, the dividing line between public and private responsibility can be blurry. For the most part, utilities provide the water to the property line and the owner is responsible after that. Even if high quality water is provided, if premise plumbing is poor, water quality and service can deteriorate. The plumbing systems are regulated through policies in the housing, social, and health arenas. Building codes provide the only real controls on premise plumbing, and these are influenced by various laws. For example, under the SDWA leaded solder has been prohibited in new plumbing systems.

Small water systems

Small water utilities must meet regulations with limited resources and capacities, but most compliance problems occur in these small utilities and development of staff capacity is of urgent importance. Small utility management can be compared to small business management or even household management. In your home, you are responsible for all the utilities, but you cannot do everything. You must call plumbers and electricians sometime. The dilemma is that small utilities might lack resources to operate independently and have to rely on others for help, such as state regulators, associations, local larger utilities, and extension agents.

WATER SUPPLY GOVERNANCE AROUND THE WORLD

Governance of the water supply sector in other countries mirrors the U.S. experience but there are significant differences. While generalizations are risky, practices vary mostly between developed and developing countries.

In developed countries, most seem to have a set of health-driven rules for drinking water, but they may be administered differently. For example, Canada has some two dozen laws related to drinking water, which are administered by Environment Canada (2010), but regulation of drinking water is left to the Provinces, in partnership with the national government. The national agency has programs for First Nations water supplies, meaning the native people.

Water supply policy in the European Union evolved separately in the different sovereign countries. The EU has a Drinking water directive of 1998 which sets quality standards for drinking water quality at the tap, requires member states to monitor drinking water quality and provide consumers with information on drinking water quality, and allows exemptions for small water supplier and certain other situations. Member states can, of course, make standards more stringent. The EC issues a report every three years. The latest one showed a lot of variation in reporting and difficulty in synthesizing the results (EC, 2007).

In Great Britain the water and sewerage industry was privatized in 1989. A Drinking Water Inspectorate was formed in 1990 to oversee water safety, and the economic regulatory structure in England and Wales is operated through the Water Services Regulation Authority (Ofwat), which is one of the national sector regulatory agencies. Ofwat has responsibilities to protect customer interests, the licensing of operators and the regulation of prices through the fixing of price caps for service provision. In France, 80% of customers are supplied by private firms. Local communities can manage water services or delegate them to private firms. The Ministry of Health has responsibility for drinking water quality. In the French water supply sector there is no specific pricing scheme for the poorest households. Rather, the approach is for financial aid to help qualified low-income households pay their water bills (United Nations Research Institute for Social Development, 2010).

The British experience in privatizing water utilities has been studied extensively.

In Latin America, each country has a different approach. Most countries have separate ministries for health and environment. The water supply sector in Brazil is dominated by regional state companies with concessions from municipalities, including large cities like São Paulo. Many municipalities still provide their own services, however, and a variety of arrangements occur across the country. The policy was dominated until the late 1980s by the National Housing Bank (BNH). Social policy focused on investments in water supply (sewage was not a priority) and a system of cross subsidies. Now, a variety of programs exists, including increase in coverage of sanitation systems, institutional and human resources development, creation of worker cooperatives, loans to low-income families, sanitation for Indian tribes, and services in rural areas. Social tariffs are widely used. In Colombia the private sector has historically participated in the water and sanitation sector both as service provider and subcontractor for the provision of infrastructure.

Japan's first modern waterworks was in Yokohama in 1887. The Waterworks Act of 1957 was the first large advance in regulation and initiated a large increase in service coverage from about 30% to nearly 100% today. This new focus on service improved water safety so people could use water without boiling. The Act provided for: drinking water quality standards, standards for facilities, licenses and supervision for water utilities, financial assistance, and other rules. Governance including business licenses is by the Ministry of Health, Labor and Welfare (MHLW). For smaller water utilities, prefecture governments are in charge. Water utilities are normally operated by municipal governments and any utility serving more than 100 requires a license. There are 9202 municipal utilities. There are also some 102 bulk water providers, in most cases managed by a prefectural government or cooperative body organized by municipalities. The MHLW licenses water utilities with the service population of 50,001 or over, or bulk water with maximum volume of more than 25,000 m^3/day. Prefectural governments license water utilities or bulk water supply for smaller entities. Water utilities may entrust services to a third party. Facility standards are performance standards so tap water can meet drinking water quality standards reliably. Drinking water quality standards are prescribed for 50 items (as of April 2009) in accordance with WHO-Guidelines for Drinking Water Quality. Financial regulation is by the municipalities with central guidance. They use enterprise accounting and a tariff system so that cost is covered by tariffs. Tariffs are a progressive system, higher rates are charged as more water is consumed (Japan Water Works Association, 2010).

WATER SUPPLY IN DEVELOPING COUNTRIES

Regulation of drinking quality in developing countries varies widely and seems to depend on the effectiveness of regulatory controls in the health and urban services arenas. Access to drinking water and sanitation in these countries has been an international development priority for decades and, as a consequence of lack of access, many people suffer disease and infant mortality is a serious problem. This issue is complex because its solution depends on the larger issues of economic, social, and institutional development.

The drinking water issue in developing countries is discussed in more detail in Chapter 12. International attention to the issue increased in 1977 when the United Nations Water Conference at Mar del Plata, Argentina concluded that just over one billion people in rural areas lacked access to a safe water supply or sanitation. Then, the UN proclaimed the 1980's as the "International Drinking Water Supply and Sanitation Decade." Progress since then has been slow, but attention remains focused on the issue.

Access to safe drinking water is a critical issue in developing countries.

The US Agency for International Development (1990) organized a project for Water and Sanitation for Health (WASH) and published a set of lessons learned about governance policies that emphasize broad participation and not just by government. WASH found that technical assistance is most successful when it helps people learn to do things for themselves in the long run. It saw that water supply and sanitation development proceeds most effectively when its various elements are linked at all levels. The most important measure for success at the national and community levels is sustainability after donor assistance has been terminated. Goals include establishing autonomy in financial issues, personnel, policies and planning; providing effective leaders and skilled managers; hiring, training, and retaining personnel; setting up administrative procedures and policies; maintaining a commercial orientation and financially sound practices; making sound technical decisions in all areas; creating a positive organizational culture; and providing quality services to consumers (Interagency Task Force, 1992).

At the global level, the World Health Organization (WHO, 2010), based in Geneva, Switzerland, is the agency within the United Nations system to provide leadership on health matters, to set norms and standards, and to provide technical support. It publishes internationally-accepted guidelines for drinking water quality. WHO also operates through regional centers, such as the Pan American Health Organization, based in Washington, D.C.

CONCLUSIONS

Effective governance of the water supply sector has the highest profile among all global water issues because of its responsibility for public health. The main commodity of the sector is treated water delivered mostly by utilities, but the sector includes self-supplied users, point-of-use treatment and distribution vendors, premise plumbing systems, and bottled water.

In developed countries, most potable water supply is provided through utilities that are regulated by safety inspectors. Publicly-owned providers are usually self-regulated for rates, but many private companies operate monopoly systems and are usually controlled by government regulators.

In developing countries, utilities may exist to provide water to higher-income urban areas, which may be served in a similar fashion to developed countries, but lower-income areas, including informal settlements, may lack access to safe water. In many rural areas (and parts of urban areas) of developing countries the water supply sector involves village-level systems, which are not be organized like utilities. People may have to take care of their own needs with individual or group wells, small community-based systems, and a variety of solutions, which at their worst may involve carrying water long distances from polluted sources.

Governance in large urban areas of all countries is challenged to develop and sustain massive networked infrastructure systems that tap water from many sources to process and deliver it safely and reliably to millions of customers. Utilities in these areas must finance, develop and operate complex enterprises to provide safe and reliable water.

With rising pressure on available water supplies, the sector is challenged to gain access to new, high quality supplies and to continue to deliver reliable and safe water through aging infrastructure systems amidst other challenges posed by its regulatory and operating environments. The sector is examining ways to provide water of different qualities, such as reclaimed water, to improve efficiency in its use.

As the water supply sector faces the future, it confronts a number of issues of concern to consumers and the public at large. In the U.S., Means (*et al.*, 2005) and Mann and Runge (2009) have published results of surveys of utilities. Results show utility executives concerned about business issues, such as the capacity to do increasingly-complex jobs with the workforces they can sustain; infrastructure breakdowns and rising water rates; access to source water supplies and worry about loss of water supply during climate change; meeting regulations; the capacity of small systems; and security against a range of threats.

To face these challenges, the sector must empower and support utilities to sustain their capacity at all levels, especially respond to the needs of lower-capacity providers; educate and empower end users; exercise vigilance to maintain effective and responsive regulatory programs; keep vibrant workforces; preventing contamination of source waters; develop adequate financing systems; and continually make advances in technologies.

Postscript

It is a joke in our family that we have a "mixed marriage." My wife's ancestors lived in the North of the U.S. and mine lived in the South. During the Civil War her ancestor (a new immigrant from England) was in the Ohio Volunteers and my mine was in Robert E. Lee's Army of Northern Virginia. That war cost more U.S. casualties than all of the other wars together, and more of them were due to sickness and disease than to battlefield losses.

My Confederate ancestor was a physician in Lee's Army. His job was to take care of the Army's physical needs such as drinking water, medicine, surgery and anything else that was needed. This was difficult because, not only was 19th Century medicine primitive, it was especially difficult in military camps. Water supply was a major problem. Bad water was a major cause of dysentery and other waterborne diseases, which accounted for much of the misery and death.

It was not only the soldier who suffered from bad water. Many civilians died from waterborne disease too. Soon, however, a bright light appeared in the form of microbiology to explain the consequences of bad water and to lead to solutions. The evolution of systems for water supply governance followed, and today waterborne disease is a much smaller problem, at least in developed countries.

Perhaps my most intensive experience with water supply governance was 12 years of service on the Fort Collins Water Board, one of the city's advisory boards. For the most part, our concerns focused on water rights, finance, local politics, water conservation, and other water quantity topics, rather than water quality. The water supply utility had competent water quality staff and they seemed to have things under control, other than a few incidents such as an outbreak of Giardia that caused some sickness. The cause of this, by the way, was traced to beavers near the water treatment plant.

The major water supply legislation in the U.S. is the Safe Drinking Water Act, which was passed in 1974. I was fortunate to be teaching about water resources prior to that and to witness the advent of the legislative program. Then, in conjunction with my studies of infrastructure management, I was able to participate in a drinking water research program that gave me much more insight into the full range of water supply issues. This is the program of the Water Research Foundation (formerly American Water Works Association Research Foundation, AWWARF), which studies all aspects of water supply improvement. Through this participation, I've become much more aware of how water supply is governed, and two aspects of it have riveted my attention.

One is how the global water supply community has developed the concept of the multiple barrier approach to protecting drinking water quality and public health. This means to protect water quality at the source, at the treatment plant, in the distribution system, and at the tap. The other aspect is the multiple challenges in maintaining a vast distribution system infrastructure that in the U.S. reaches some two million miles, not counting the premise plumbing.

Think of all the governance issues involved in these multiple phases of water supply and infrastructure challenges. The water supply community has its work cut out for it, but it comprises many talented and dedicated men and women who are up to the task if they are given the resources and freedom to act in the public interest. I feel blessed to continue to work with them in water supply research and improvement.

REFERENCES

Colorado Department of Public Health and Environment (2009). *Safe Drinking Water Program Capacity Building Unit*. http://www.cdphe.state.co.us (accessed 4 January 2010).

EC (2007). *The Quality of Drinking Water in the European Union 2002–2004*. http://circa.europa.eu/Public/irc/env/drinking_water_rev/library?l=/drinking_synthesis/report_2002-2004pdf/_EN_1.0_&a=d/ (accessed 7 January 2010).

Environment Canada (2010). *Comprehensive Approach to Clean Water*. http://www.ec.gc.ca (accessed 2 April 2010).

Grigg N. S. and Zenzen M. (2009). *Water and Wastewater Workforce Planning, Design, and Action for Organizational Excellence*. AWWA Press, Denver.

Japan Ministry of Health, Labour and Welfare (2010). *Policy and Administration of Water Supply in Japan*. www.mhlw.go.jp/english/policy/health/water_supply/5-4.html (accessed 6 January 2010).

Japan Water Works Association (2010). *The Present Condition of the Water Service in Japan*. http://www.jwwa.or.jp/english/water_en/water-e07.html (accessed 28 May 2010).

Mann J. and Runge J. (2009). State of the industry report 2009. *J. Am. Water Works Assoc.*, **101**(10), 36–46.

Means III E. G., Ospina L. and Patrick R. (2005). Ten top trends and their implications for water utilities. *J. Am. Water Works Assoc.*, **97**(7), 64–75.

National Ground Water Association (2010). *Informing Consumers about Water Wells*. http://www.wellowner.org/ (accessed 4 January 2010).

United Nations Research Institute for Social Development (2006). *Social Policy, Regulation and Private Sector Involvement in Water Supply*. http://www.unrisd.org (accessed 5 January 2010).

USEPA (2008). *Factoids: Drinking Water and Ground Water Statistics for 2008*. http://www.epa.gov/safewater/data (accessed 4 January 2010).

Water Research Foundation (2010). *Strategic Research Goal Areas*. http://www.waterresearchfoundation.org (accessed 4 January 2010).

World Health Organization (2010). *Guidelines for Drinking-Water Quality*, 3rd edn. http://www.who.int (accessed 15 June 2010).

Chapter 6

Wastewater and water quality management

While water supply is the most critical water service, the wastewater and water quality sector may be more complex and difficult to govern. It deals with three distinct issues: access to wastewater service, management of wastewater utility services, and environmental water quality. The sector deals with water quality from "cradle-to-grave," so to speak. Unfortunately, many water quality issues are complex and difficult to control within the confines of organized treatment systems.

Access to sanitation and wastewater services is a companion public health issue to safe drinking water and needs attention in many parts of the world. Wastewater treatment is an expensive capital- and energy-intensive service and has not been widely adopted everywhere. Nonpoint sources of pollution are the most difficult to manage. The vast and sprawling wastewater and water quality management sector is a challenge to govern and even to understand. However, the sustainability of environmental water systems depends on effective governance of it.

The wastewater and water quality sector deals with water quality from "cradle-to-grave" to safeguard public health and protect the environment.

WASTEWATER AND WATER QUALITY: BASIC NEEDS OF HUMANS AND THE ENVIRONMENT

Although it is common to refer to "water and sanitation" as linked service areas, they are quite different. WHO (2010) explained sanitation as "… the provision of facilities and services for the safe disposal of human urine and faeces" for the "maintenance of hygienic conditions, through services such as garbage collection and wastewater disposal." The linkage is that both water supply and sanitation involve protection of public health, but other than that they are different services in many ways.

Without safe and convenient ways to dispose of human wastes, people become exposed to health hazards and the consequences manifest themselves as diseases such as dysentery, cholera, and typhoid, among others that have afflicted humans from the beginning of time. Today, these are a major problem in developing countries, where around a billion people lack access to safe and effective sanitation services. Given its linkage of to poverty and development issues, access to safe water and sanitation is explained in more detail in Chapter 12.

Whereas access to sanitation services can be provided by individual and on-site systems, the management of wastewater is a utility-type service that involves collecting the wastewater, processing it, and disposing of it. Wastewater utilities normally operate networks of collection sewers that receive wastewater from buildings, and they own and operate complex and expensive treatment and disposal works. This service is complex and expensive and only seems to work well in some high-income countries.

The need for management of environmental water quality is caused by wastewater discharges and polluted runoff from a range of land uses. In developed countries wastewater discharges may seem under control, but it is very difficult to address the problem of polluted runoff. Contrary to the impression that some people have, many issues of wastewater discharges still must be addressed as well.

In developing countries, stream pollution is a result of both point and nonpoint pollution. In some cases, the pollution loads are extremely heavy due to lack of any organized wastewater management systems.

In the complex wastewater and water quality sector, not only is access to sanitation an important worldwide issue but even when wastewater service is established, it is difficult to finance and manage it effectively. When you add in the environmental problem of non-point sources, the full complexity of the sector becomes clear.

This chapter addresses governance of the wastewater and water quality sector to cover services and regulatory controls. The issue of access to service is embedded in the overall wastewater topic, but it extends to governance in the basic housing and social support sectors as much as it does to the water sector. The central focus of the chapter is on utility and land use systems and their effects and interactions with water quality in the environment.

SECTOR CHARACTERIZATION

The main situations where wastewater and water quality governance are applied are shown on Figure 6.1, which illustrates point and nonpoint source dischargers and the need to manage environmental water quality. On the figure can be seen municipal and industrial point source dischargers and several categories of non-point sources.

The three distinct service categories (access to sanitation, wastewater treatment, and environmental water management) are shown in Figure 6.2, which illustrates where they occur in the water management cycle. Access to sanitation is at the premise plumbing locations, and the generation of polluted runoff is shown at the drainage systems. Three streams of used water enter the receiving water: point discharges, runoff from individual systems, and runoff from stormwater after any best management practices (BMPs) have been applied.

The two drivers of demand for wastewater management are wastewater disposal and land use. Human–generated wastewater adds contaminants to water that range from body wastes through diverse industrial effluents. Waste products from land uses fall into the nonpoint source categories, which will be explained later. Human-generated wastes are mostly collected as wastewater in organized sewerage systems, whereas land use adds waste materials to runoff and leads to the non-point source problem.

The first box shows the wastewater utility, which manages collection, treatment, and disposal. For the most part, these generate point sources to receiving waters. Publicly-owned systems such as collection sewers are under management of the utilities and premise plumbing is owned by and the responsibility of individuals. Wastewater management for residential, commercial, and industrial customers is normally by utilities, but many industries and commercial sites have their own separate disposal units.

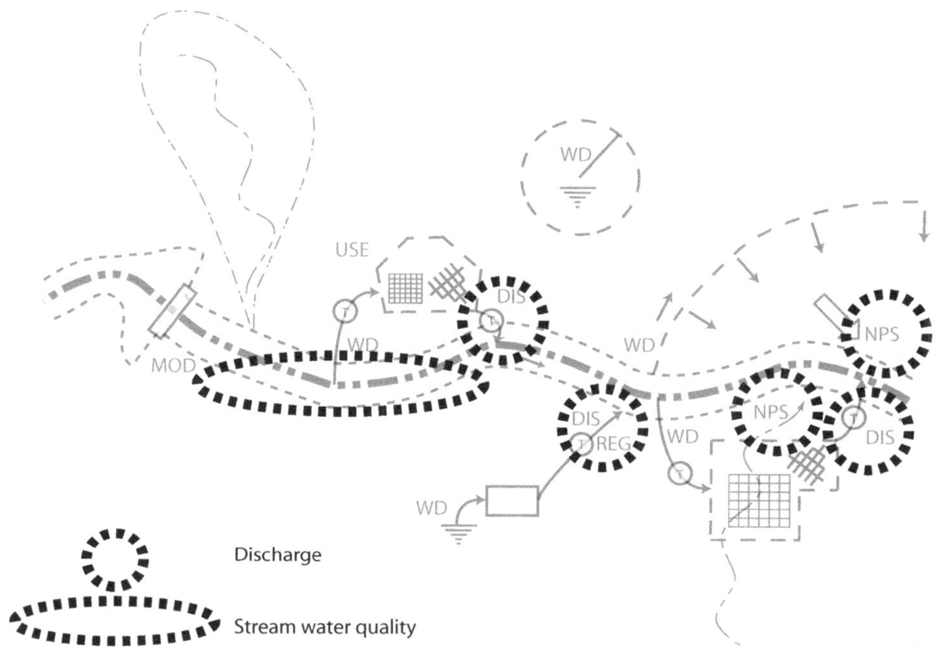

Figure 6.1 Discharge and stream water quality control points

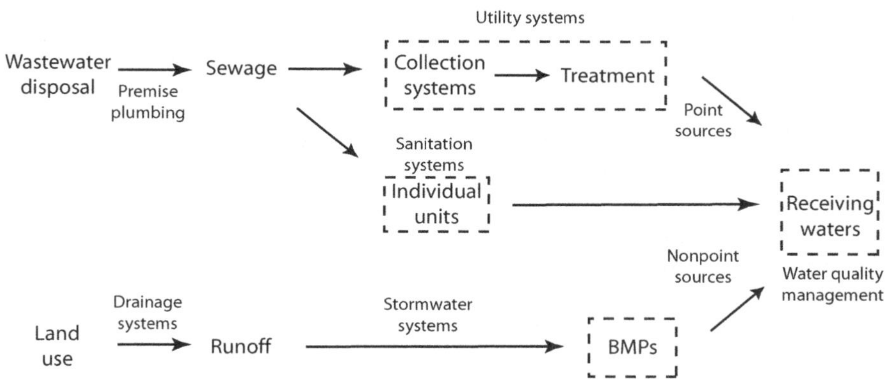

Figure 6.2 Wastewater, sanitation, and water quality processes

The second box shows individual disposal units, which provide rural and non-utility services. In the U.S., most individual dischargers require permits, and these vary from the smallest discharger using a small septic tank to larger systems, such as a community collection system going to a disposal tank and treatment works. Individual units are normally privately-owned, but a community system might be provided and managed by a homeowner's association or even a public utility. These systems usually do not discharge as point sources but rely on disposal units such as septic tanks. Much of the wastewater disposal in developing countries is of this type rather than in piped collection systems. This occurs because waste disposal in villages or cities which lack sewers use individual units or no units at all.

The third box comprises the best management practices of the non-point source control component. Non-point sources of pollutants may or may not have permits. The Clean Water Act has spawned complex rules about permitting stormwater control.

The fourth box shows the environmental water quality control activity. It comprises a range of activities from monitoring and assessment through reporting and policy. Environmental water quality is monitored, and feedback actions are used to regulate the dischargers.

GOVERNANCE OF THE SECTOR

Policy and goal-setting

Policy for the wastewater and water quality sector must deal with access to service, managing wastewater systems, and environmental water quality.

Similar to water supply, access to wastewater service involves social and health policy, and has interfaces with housing where controls such as building and plumbing codes are used to set standards. Household facilities are provided mostly by private owners, along with agencies charged with providing affordable housing.

In developed countries, access to sanitation is well-established, but many other problems must be faced.

In the U.S., the availability of sink, toilet, and bathing facilities is practically universal, but the sanitation issue involves many concerns about public education, maintenance of facilities, and other public health issues. Access to service in public and commercial places requires continuing vigilance, and is controlled by diverse rules and regulations. The U.S. has no regulatory program for wastewater service comparable to its regulation of "non-community water systems" (see Chapter 5).

Access to service in developing countries is an urgent issue when living conditions are so poor that people simply cannot access adequate waste disposal facilities. This is the problem behind the global sanitation issue that is explained further in Chapter 12. Many different types of systems are in place but without regulatory controls in most places.

Wastewater management is a utility service and involves large investments in infrastructure, management of organizations, and compliance with regulatory controls. In the U.S., wastewater service in cities was mostly implemented in the late 19th Century in response to demands of urban residents (Melosi, 2008). Household plumbing systems brought water supply and wastewater disposal services and made life cleaner and more convenient for city dwellers. By the 1950s, most of the U.S. had sewer networks for wastewater disposal, but it had few treatment plants and most governance actions were by state and local governments, not at the national level.

Although the federal government had addressed wastewater in some legislation, firm policy for the sector only emerged with the 1972 Clean Water Act (CWA), which was passed during a period of intense concentration on environmental problems with emotions fanned by a famous scene of a polluted river catching on fire. The CWA addresses wastewater management from the standpoints of both public health and environmental water quality.

The Clean Water Act is the fundamental U.S. law for environmental water quality.

In passing the CWA, the U.S. debated whether to implement a system for regulation or economic incentives for water quality control (Grigg and Fleming, 1980). The debate was resolved with regulatory controls over wastewater treatment, nonpoint source control and protection of stream water quality. While USEPA does not regulate the integrity of collection systems directly, it administers a program of assistance for them.

Strong incentives work against wastewater service being established as a self-supporting utility service because people resist paying the full cost of treating wastewater that leaves their premises. This accounts for some of the difficulty in establishing wastewater treatment and in sustaining environmentally-responsible systems.

Environmental water quality management is managed indirectly through the CWA by placing controls on discharges to streams and requiring monitoring and reporting. Although the system requires continuing vigilance, the U.S. has an effective system of controls on point source dischargers, the difficult part is in implementing effective non-point source controls.

For example, when the CWA was enacted, stormwater policy was non-existent. USEPA approached policy development in phases. Now, the policy is to regulate runoff from Municipal Separate Storm Sewer Systems (MS4s) through permit programs. The rule for Phase I, which was issued in 1990, required larger urban areas to obtain NPDES permits for their stormwater discharges. Phase II, which was issued in 1999, extended the permit program to smaller areas. For the most part, the larger systems have individual permits and the smaller ones fall under general permits. Each regulated system must implement a stormwater management program to reduce the stormwater runoff contamination and prohibit illicit discharges (USEPA, 2010a).

Another nonpoint source issue is national policy on estuary cleanups, which was established in 1987 via a National Estuary Program through an amendment to the Clean Water Act. It veered away from the traditional regulatory approach to engage local communities to take responsibility for managing their estuaries through a coordinated approach involving government and community stakeholders.

Now USEPA is implementing the "Total Maximum Daily Load" or TMDL program to assign carrying capacities to streams and to allocate permission to discharge wastes. The difficult part is, once again, how to handle the non-point sources.

Empowerment

The need for empowerment in the wastewater sector arises from three forces: difficulty experienced by lower income people to get access to safe and adequate facilities, incentives that work against establishment of effective wastewater utilities, and resistance to adequate funding for the wastewater field as a target for investment.

Empowerment for access is about solving the problems of poverty and provision of adequate services and living conditions in all public places and dwellings. The problem of access to water supply and sanitation in developing countries was discussed briefly in Chapter 5 and how empowerment relates to poverty is discussed in Chapter 12. This explains the high levels of attention given to the sanitation issue by international donors and government and non-governmental organizations. One solution is, of course, a rising standard of living, which might solve the problem of access to service.

Empowerment in wastewater treatment and environmental water quality requires direct policy intervention because of the lack of incentives for people to take care of wastewater and solve nonpoint source problems. These require continuing inputs for technical and financial assistance and capacity-building.

Empowerment in wastewater treatment must address the issue that smaller management organizations may be overwhelmed by the complexity and cost of the programs. Two common areas of empowerment are financial assistance and training for operators. Another approach is on a continuing program of research.

Empowerment to enable the function of environmental water quality management is required to address issues of institutional capacity. A management organization will require staff, resources, training, and authority to mount monitoring, reporting, data management, and enforcement programs.

Control

The control function of governance is focused on wastewater management, where the quality of discharges is regulated, and on environmental water quality management, where the quality of the water depends on success in regulating point and nonpoint sources. Most of the scenarios described in the next section are about controls.

SCENARIOS OF WASTEWATER AND WATER QUALITY MANAGEMENT

The following scenarios were selected to illustrate the range of governance issues that occur in the diverse subfields of wastewater and water quality management. For most of these I had some experience in administering programs at the state government level, and some have been added to fill out the range.

Regional water quality planning

Environmental water quality is a shared concern of everyone in a watershed and, given that it results from both point and nonpoint discharges, it requires monitoring and study to know what to do. The CWA contains a provision for areawide wastewater management planning under Section 208 where the idea was to derive watershed solution strategies that consider both point and nonpoint sources. Upon implementation of the Act, the states developed programs for the 208 studies and published comprehensive plans based on water quality analysis. These plans served mostly for problem identification and required additional studies and planning later.

Section 201 of the CWA serves as a control mechanism in providing for coordination of wastewater treatment planning in regions. As originally implemented, this requirement was linked to receipt of financial aid through the Construction Grants Program, so there was a strong incentive for utilities to comply. This had the result that utilities were not able to apply for grants without subjecting their plans to coordination with other wastewater treatment providers in the same areas.

Now, most states have good ideas of the basic issues within watersheds and are concentrating on nonpoint source controls and on Total Maximum Daily Load (TMDL) studies.

Total maximum daily load program

When Congress passed the CWA, it included Section 303(d) that required states to identify waters that were limited by water quality. Then, the states were to establish priority rankings and TMDLs. The requirement was ignored for many years, but a court case (Sierra Club vs. Hankinson, 1996) led to a consent decree and forced USEPA (2010b) and the states to develop TMDLs.

A TMDL is a calculation of the maximum amount of a pollutant that a waterbody can receive and still meet water quality standards and support designated uses. It comprises wasteload allocation of point sources, a load allocation of nonpoint sources, a margin of safety, and consideration of seasonal variation.

TMDLs are supposed to be used to allocate loads equitably between all sources in a watershed and to restrict new or expanded discharges. TMDL studies include watershed characterization, data analysis, methodology selection, data collection, model calibration, scenario development, and implementation plans.

The TMDL program addresses the water commons from the standpoint of water quality and involves complex calculations and the potential for conflict. In a sense it is the companion issue to instream flows for water quantity, which are addressed in Chapter 8.

Discharge permits

Although nonpoint sources are important and difficult, regulating point source discharges from cities and industries remains the first line of defense against water pollution. Because the CWA is basically a regulatory program, the control system for point sources is based on the discharge permit system, where all facilities that discharge into receiving water require National Pollutant Discharge Elimination System (NPDES) permits (USEPA, 2009). This permit program is the core strategy for regulation of point source dischargers in the U.S. It is a massive program, and more than 400,000 facilities were required to have NPDES permits by 2001, and the number was expected to grow to more than 500,000 within a few years (USEPA, 2001). The inventory of permits includes some 16,000 public wastewater treatment plants, large and small industries, military bases, schools and universities, and other types of facilities.

NPDES program uses a "command-and-control" regulatory approach, but the U.S. could have implemented a program of economic control of point source discharges. In this type of program, a city or industry would be charged an amount equivalent to the level of pollutants discharges. Sometimes this is called a "polluter pays" program.

Nonpoint sources

Nonpoint sources are characterized by USEPA as "polluted runoff" and controlled through different programs than point source discharges. Categories are: agriculture, forestry, habitat alteration, marinas and boating, mining, roads, highways, and bridges, urban runoff, and wetland or riparian management. For the most part, these are covered by best management practices (BMPs) but in some cases, permits are required under the stormwater program. Examples where permits are required are municipal stormwater systems, industrial stormwater, and construction sites over five acres. In 2009, EPA issued a rule requiring all construction sites of one acre or more to use BMPs to control sediment.

Empowerment through financial assistance

Once people learn that their health depends on safety of their drinking water, they are willing to spend money to obtain clean water However, they will resist paying to treat wastewater because the benefits impact many other people, rather than the payers. Recognizing this reality, the U.S. made massive investments during the 1970s in a system of wastewater treatment under the Clean Water Act. This had the effect of jump-starting the wastewater program, but now the nation has a legacy of wastewater treatment assets that require a lot of maintenance and operating expenses.

Financial assistance was included in the CWA through the Construction Grants program, which financed some $60 billion in new publicly-owned treatment plants, pumping stations, collection and interceptor sewers, sewer rehabilitation, and combined sewer overflow controls in the 1970s and 1980s. The 1987 amendments to the Clean Water Act replaced the program as of 1990 and initiated a Clean Water State Revolving Fund (CWSRF) (USEPA, 2010c). From its enactment through 2010, the CWSRF program has funded over $68 billion through some 22,700 low-interest loans for wastewater treatment, nonpoint source pollution control, and watershed and estuary management.

Operator training and certification

Wastewater treatment plants are complex, and once they are in place, it is difficult to operate and maintain them for long term effectiveness. This requires certified operators who are trained and supported and the U.S. has nearly 40,000 wastewater operators (Grigg and Zenzen, 2008). This presents a special workforce challenge for smaller systems, and the CWA provided for financing of pilot programs for training and development, such as the Wastewater Treatment Plant Operator on-site assistance training program at small publicly-owned plants.

Federal funding flows through states and assistance is administered by environmental training centers around the country (USEPA, 2010d).

Individual systems

Control of the use of individual on-site systems is normally regulated by state and local governments, which impose requirements for construction standards. If these units operate properly, the do not discharge significant quantities of pollutants to receiving waters, but if they are not designed or maintained well, they can become heavy polluters and the runoff from them is regulated through the nonpoint source program for water quality control.

Environmental water quality

Environmental water quality is controlled by a system of stream standards and ambient monitoring. It results from success in regulating point and nonpoint sources through other programs. Planning and coordinating ambient stream water quality is an important control mechanism because permits and controls on nonpoint sources seek minimum standards, and sustainable water management requires better performance than that. For that reason, the TMDL program provides an additional mechanism to assess the effectiveness of overall regulatory controls. Another control mechanism is the required bi-annual reporting by states of the condition of their streams. This is handled under Sections 305(b) and 303(d) of the Clean Water Act.

Enforcement

A water pollution regulatory program must include an effective enforcement arm. In the U.S., basic enforcement is at the state level, and overseen by USEPA which signals its importance by assigning compliance and enforcement to a separate department at the level of an assistant administrator.

The theory of any form of law enforcement is to assign a graduated series of penalties to correspond to varying offenses and to achieve compliance with the law. As an example of a minor offense, if a utility was late filing a required report of some kind, it might receive a letter of reprimand. If it flagrantly violated its permit condition, it might receive a civil penalty amounting to the money saved by not complying, for example. If officials violated permit conditions and concealed the facts, they might go to jail under criminal law.

USEPA describes enforcement results by two measures, the estimated pounds of pollutants reduced and estimated cost to defendants to control or reduce pollution. In fiscal year 2009, enforcement actions were taken at about 4,600 facilities. The civil and criminal actions required polluters to invest more than $5 billion on pollution controls, cleanup, and environmental projects. The results of enforcement are measured in pounds of pollutants reduced, and the estimate was that FY 2009 actions would remove about 580 million pounds of air, water, and land pollutants annually once required controls are implemented.

Voluntary EMS/ISO 14001

Wastewater plants offer a good example of how voluntary environmental quality programs with integrated goals can be used. USEPA is promoting Environmental Management Systems (EMS) for communities because it sees them as tools to improve environmental performance, implement pollution prevention, and improve compliance. This might increase regulatory flexibility for organizations that use EMS, if they are successful. USEPA's Offices of Water and Compliance are working together to provide training and technical assistance to local governments to implement ISO 14001, the International Organization for Standardization program for voluntary environmental standards. This is part of the ISO 14000 series, which includes environmental auditing, environmental performance evaluation, environmental labeling, and life-cycle assessment. ISO 14001 is designed to help communities and other organizations meet environmental obligations and reduce their impacts on the environment.

Virginia Tech University (2010) has an EMS institute in place to train local officials in implementing plans, including at wastewater treatment plants. The institute provides cross-program training to assist local officials in improving compliance and reducing pollution from activities such as stormwater runoff, air emissions, control of solid and hazardous wastes, and other operations at a wastewater treatment plant.

WASTEWATER MANAGEMENT IN OTHER COUNTRIES

As is the case with water supply, wastewater policy is similar across developed countries, although there are important differences.

For example, the E.U.'s Water Framework Directive (WFD) is its basic policy toward water pollution control and, like the environmental water quality program of the US, it focuses on integrated management in river basins (Chapter 10). A main difference is that in the U.S., states are obligated to comply with federal law, whereas the E.U. is still developing its compliance mechanisms among member states.

The E.U. wastewater directive dates from 1991 and deals with collection, treatment and discharge of urban wastewater and wastewater from some industrial sectors. The timetable started with a 1998 deadline for all areas of more than 10,000 population equivalents discharging water into sensitive areas to have tertiary treatment. By the end of 2000 areas of more than 15,000 population equivalents outside sensitive areas had to have secondary treatment, and by the end of 2005 additional requirements were imposed on areas of between 2,000 and 15,000 population equivalents, mainly to have collection and treatment systems. The most recent implementation report (as of 2004) noted that none of the deadlines has been met by all member countries (Commission of the European Communities. 2007). This is not unlike the U.S. experience, where the CWA set a high bar for goals and progress has been slower than envisioned, but still positive.

The E.U.'s Water Framework Directive outlines water quality policy for member states.

Japan has a more centralized government than the U.S. and decisions are made in a different political and cultural environment. While Japan has a Ministry of the Environment, water pollution control activities are worked out on a more cooperative basis than in the U.S., which tends toward an adversarial approach.

Basic water quality policy in Japan is by the national government, and the enforcement system is operated by prefectural and municipal governments. There is some financial assistance to designated facilities, tax incentives, and public investment in urban wastewater systems (Imura, 2004). The timeline for development of wastewater programs was similar in some ways to the U.S., with modern systems emerging mostly after about 1950. The 1958 Sewerage law was a precursor for later, more comprehensive approaches, such as the 1962 decision to place wastewater management in the Ministry of Construction (now the Ministry of Land, Infrastructure and Transport) and the first five-year program for wastewater construction followed in 1963. The Japan Sewage Works Association was established in 1964, and in 1970 a water pollution law was promulgated. By 2000, the percentage of population with sewer service reached 50%.

The water quality monitoring system is operated by the Ministry of Land, Infrastructure and Transport, which also coordinates a program to designate "specified facilities, such as factories and commercial facilities (hotels, agriculture, cleaners, etc.) and the operators of large facilities (13% have more than $50\,m^3$/day) and 87% have (less than $50\,m^3$/day).

The Minister of the Environment issues policy for areawide pollution load control for designated water bodies, the prefecture governors implement plans for targets and measures and use regulation, guidance, and project development for improvement of sewerage systems. The prefectures and some designated cities engage in enforcement by requesting reports from specified facilities, conducting on-site compliance inspections, and taking administrative action such as improvement orders.

As an example of enforcement levels, in 2002, there were some 300,000 specified facilities. In that year, there were 55,332 on-site inspection visits, 8,434 administrative guidance orders, and 40 improvement orders.

In developing countries, access to sanitation services is the highest-profile issue, but their decisions to implement wastewater treatment are also important. For example, in Colombia only about 5% of municipal wastewater was treated (as of 2004) and the extent of industrial discharges to sewer networks was unknown. The promotion of wastewater treatment plants has lagged due to inadequate funding. Even after political leaders become convinced of the importance of wastewater treatment, it will be difficult to generate the political will to pay for them. Financial capacity is highly variable around the country, and the system of fixed treatment requirements and lack of subsidies has worked against solving the problem. Many communities lack the income to pay for modern and effective wastewater treatment systems, even to pay for operations and maintenance of plants.

CONCLUSION

In the U.S. and other developed countries, wastewater service has changed from a low-profile municipal program of providing sewerage to a more dynamic utility program, with complex infrastructure, enterprise financing, and stringent regulatory controls. In developing countries, access to sanitation is a principal issue, and wastewater treatment is lagging almost everywhere.

Globally, the wastewater and water quality sector faces difficult challenges that include providing infrastructure systems for billions of urban residents, developing basic systems for populations who cannot afford higher-standard infrastructure systems, maintaining vast networks of aging buried infrastructure and tens of thousands of treatment plants, and paying for the ongoing operation of the systems.

In contrast to the water supply sector, where policies are similar across a range of nations, there is more divergence in wastewater and water quality policies due to different stages of national development and the fact that wastewater and water quality sector is not viewed with the same urgency as water supply. The result is that, while you find similar wastewater treatment programs across a few developed countries, there are wide variations among most and the financial crisis will exacerbate the differences.

You will also find some similar approaches to environmental water quality management such as in the U.S. and the policy of the European Union, but some countries lack any meaningful program at all. Without the management infrastructure for wastewater treatment and nonpoint source control programs, it is not possible to gain much control over environmental water quality.

As a result of the wide variations in wastewater and water quality management across the globe, this sector demonstrates clearly how governance depends on much more than regulation alone. Even in the U.S., it was not always a given that wastewater service would be developed by municipalities. Implementing effective services required both a carrot and a stick on the part of the federal government to get the sector to the point it is today. This came in the form of the 1972 Clean Water Act. The E.U. has passed rules, but member states are still coming into compliance.

Postscript

Back in the 1950s my baptism into environmentalism was by fishing on the Alabama River. We set lines mainly for catfish in a sluggish southern stream with lots of tributaries, one of which was named Catoma Creek after a local Indian tribe. Catoma Creek drained the city's sewerage system, which had a collection system but no wastewater treatment. By the time the sewage reached our fishing grounds, the river had diluted it and we hardly noticed. It was an example of "the solution to pollution is dilution." However, this was not a full solution. We just accepted it as we did not know any other way.

If you went on Catoma Creek itself, the pollution was so gross you could not stand it. No one went there or fished it. For all practical purposes, it was a dead creek. A shame. During the 1950s the city built its first primary treatment plant and now it has advanced treatment and meets Clean Water Act standards. Things have improved, as they in many places around the world. Sadly, there are still too many Catoma Creeks out there, and you can see many pictures of them in cities and built areas of developing countries.

I didn't realize it at the time, but I was to experience a historic era in the U.S. of dealing with water pollution. When I entered the consulting engineering business in the 1960s, my partner, Dave Sellards, explained to me the work done in stream sanitation by Clarence Velz at the School of Public Health at the University of Michigan. Velz wrote a book that opened my eyes to oxygen in streams and other environmental water quality issues. When I took over oversight of water quality regulation in the North Carolina State government in 1979, it was like I met the Clean Water Act all at once. We had responsibility for water quality planning, permits, enforcement, and construction grants. Daily work to administer these complex programs consumed us with many technical and political matters.

Looking back at the governance issues we faced, I am impressed by how many people were involved, how the policy was multi-level, how the incentives played in different ways among the players and some helped while others hurt the programs. Overall, I gained a lot of appreciation for the staff in the water quality program, who had to do the studies, issue the permits, monitor the streams, and enforce the regulations.

The utilities and small dischargers play the major part, of course. Large, urban utilities had to build vast infrastructure and operate complex plants and programs. Smaller dischargers struggled, especially "mom and pop" industries, of which there are many in North Carolina. The whole program required a steady hand on governance to make sense and to work out. Today, some 30 years later, it seems to me that the system is working, but it requires a lot of vigilance to be sustained effectively. Many countries have not reached the programmatic levels that I described above. In some of their cities, people even lack access to wastewater collection systems. A lot of work is needed to provide effective wastewater and environmental water quality systems around the world.

REFERENCES

Commission of the European Communities (2007). *Towards Sustainable Water Management in the European Union: First Stage in the Implementation of the Water Framework Directive 2000/60/EC.* http://ec.europa.eu/environment/water/water-framework/implrep2007/pdf/sec_2007_0362_en.pdf (accessed 7 January 2010).

Grigg N. S. and Fleming G. (1980). River basin management: U.S. national experience. In: *International Association for Water Pollution Research Specialized Conference*, June, Cincinnati, OH.

Grigg N. S. and Zenzen M. (2009). *Water and Wastewater Workforce Planning, Design, and Action for Organizational Excellence*. AWWA Press, Denver.

Imura H. (2004). Optimizing compliance monitoring and enforcement: Japanese experience in water pollution control. In: *OECD Global Forum on Sustainable Development*, Paris.

Melosi M. V. (2008). *The Sanitary City: Environmental Services in Urban America from Colonial Times to the Present*, 2nd edn. University of Pittsburgh Press, Pittsburgh.

USEPA (2001). *Protecting the Nation's Waters Through Effective NPDES Permits. A Strategic Plan: 2001 and Beyond.* http://www.epa.gov/npdes/pubs/strategicplan.pdf (accessed 17 October).

USEPA (2009). *Water Permitting 101*. http://www.epa.gov/npdes/pubs/101pape.pdf (accessed 26 November 2009).

USEPA (2010a). *Stormwater Discharges From Municipal Separate Storm Sewer Systems (MS4s).* http://cfpub.epa.gov/npdes/stormwater/munic.cfm (accessed 8 January 2010).

USEPA (2010b). *Impaired Waters and Total Maximum Daily Loads.* http://www.epa.gov/owow/tmdl/ (accessed 29 May 2010).

USEPA (2010c). *Clean Water Financing*. http://www.epa.gov/owm/cwfinance/index.htm (accessed 8 January 2010).
USEPA (2010d). *Wastewater Treatment Plant Operator On-Site Assistance Training Program / 104(g)(1)*. http://www.epa.gov/owm/mab/smcomm/104g/ (accessed 8 January 2010).
Virginia Tech University (2010). *PEER Local Resource Center.* http://www.cota.vt.edu/vtems/index.html (accessed 18 June).
WHO (2010). *Sanitation*. http://www.who.int/topics/sanitation/en/ (accessed 6 January 2010).

Chapter 7

Irrigation and drainage

Irrigation is a large water-using sector with a unique set of governance challenges due to its close links to food production and rural development. In addition to supporting large farm enterprises, it is important to subsistence farming, which is critical to many local economies and societies. Globally, irrigation accounts for 70% of total water withdrawals, far outstripping domestic, industrial, and energy production uses. Where it is practiced on a large scale, irrigation water use dwarfs urban uses. Even in humid areas, irrigation can be important for supplemental water to improve crop yields.

Irrigation water users range from small plot holders to giant multi-national corporate farmers. Sometimes they act alone in withdrawing and using water, but more often they band together to form associations and other forms of mutually cooperative enterprises. They are not always able to finance large scale infrastructure systems solely through revenues of farming, but varied forms of managed systems come together through user charges, government subsidies, and cooperation with others in multi-purpose projects.

Given its large impacts, value to food producers and variety of system types, irrigation presents a challenge to governance. When its financial underpinnings are in many cases shaky, does it make sense to practice formal regulation? How far should governments go in subsidizing irrigation? How can its environmental impacts be mitigated? These are some of the governance questions addressed in this chapter.

Globally irrigation accounts for 70% of water diversions.

IRRIGATION AS A WATER MANAGEMENT SECTOR

World food demands are rising and irrigation is needed to sustain production. In many regions, subsistence farming relies on access to irrigation water supplies. Although irrigation places stress on ecosystems, unless irrigation water is provided, hunger will increase. As global population increases and protection of ecosystems becomes even more challenging, governance of irrigation water use will rise in the public consciousness.

While irrigation is important in the overall scheme of water management, its governance is distinctly different from urban water supply and wastewater. Water supply and wastewater are governed by utilities and regulators who focus on health and environmental issues. Irrigation is governed by ditch companies and irrigation districts and regulated mostly for quantity of withdrawals. Also, it is supported by diverse types of agricultural interest groups.

Policymakers will naturally want to facilitate access to irrigation water, regulate withdrawals to assure equitable access and environmental integrity, and regulate the quality of irrigation return flows. On the other hand, they want to respect the need for farmers to be independent and to be able to respond to market conditions as they plan their use of water.

If irrigation enterprises were like business corporations or municipalities, they could be treated like other water users and dischargers. However, irrigation water users range in size from small family farms to giant districts that practically use entire rivers for their water supplies. Another issue that distinguishes irrigators from other water users is their inability in many places to pay the full costs of water. These two attributes of irrigation—its importance to farm enterprises but their inability to pay full water costs—places special requirements on irrigation water governance.

As we look to the future, governance of the sector must confront economic and environmental issues, such as making water available, controlling use and efficiency, and managing the quality of farm runoff. It must also take into account the social objectives of empowering farmers at different levels.

Given the importance and distinct attributes of irrigation, the chapter addresses the following aspects of its governance:

- Facilitating equitable access to irrigation water and regulating irrigation water withdrawals at different levels
- Governance and management of irrigation companies
- Regulation of the quality of farm drainage

THE SECTOR

Irrigation and drainage systems divert water for agricultural use and provide for drainage of excess water back to water courses. Drainage of excess water may also occur in areas of rain-fed agriculture where supplemental irrigation is not practiced. Figure 7.1 illustrates some of the types of irrigation systems (Buyalski *et al.*, 1991).

As you can see, many different situations are possible. Starting at the top of the diagram, you see the main storage reservoir and just below the dam two main canals leave to distribute irrigation water. The main canal on the left of the diagram feeds an uphill pressure pipe system and on the downhill side it feeds a gravity system. Water is also furnished to municipal and industrial users along the way.

The main canal on the right feeds a distribution system and downstream demands. A cross-drainage canal runes over its top, which is made possible by an inverted siphon that conveys water under the drainage canal.

The natural channel leads on down the valley to a diversion dam, which feeds a third main canal's headworks. This main canal serves a center pivot system and a flood irrigation system near its beginning. Further down it serves a sprinkler system, an automated pressure pipe system, and a flash flood system. These are only a few of the system types that will occur in large irrigated areas.

The supply of irrigation water may be provided by management organizations or by individual farmers. Large scale irrigators include those who operate their own systems or take water from a large system, and smaller irrigators may have wells, small individual systems, or be connected to organized delivery systems. The dividing line between large and small is not fixed, but required water quantities can be large for even modest-sized farm plots. As an example, irrigation of a single 1000-acre farm at 2.5 feet per season would require a flow rate of about 50 cfs on a one day per week schedule.

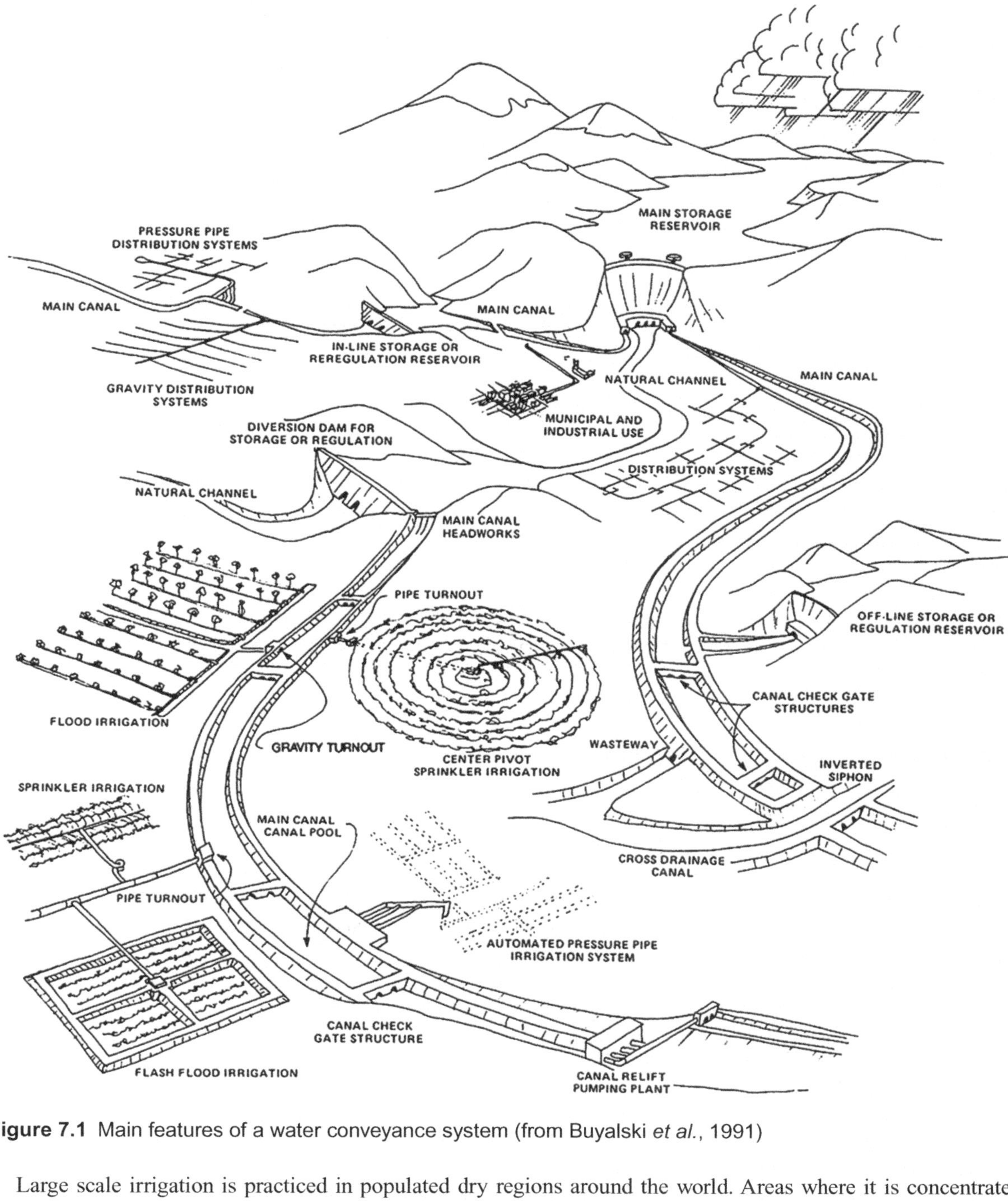

Figure 7.1 Main features of a water conveyance system (from Buyalski *et al.*, 1991)

Large scale irrigation is practiced in populated dry regions around the world. Areas where it is concentrated include the Middle East, Central Asia, the western United States, and South and East Asia (Molden, 2007). In other regions, irrigation may be used for supplemental water supply, and agricultural drainage is important everywhere. Some irrigation water uses also occur in urban water supply services and urban farming is increasing.

According to the USGS (2008), U.S. irrigation withdrawals in 2005 were about 143 MAF/year, or about 2.4 feet for some 60 million irrigated acres (out of a total of some 440 million acres of cropland). This water use is about 8 percent less than in 2000 and near the level in 1970. Use of sprinkler and micro-irrigation systems continues to increase and was 56 percent of the total acreage irrigated in 2005.

This trend toward lower total irrigation use masks shifts of types of water application. Some humid states such as Florida are increasing their application of irrigation water, and micro-irrigation is increasing in application, having reached more than 10% of all irrigated acreage in California, for example. Local officials report that outdoor water use is their most difficult problem, and it has many social and economic linkages. Thus, this problem needs continued attention.

Irrigation is increasing in cities, for landscaping, and for recreational purposes thus changing the playing field for outdoor water management.

Among the world's irrigated regions, many diverse examples of farms systems occur. In the U.S., examples of larger irrigators include large farms with surface irrigation systems for crops such as corn, alfalfa, or vegetables; large farms with wells and center pivot systems; golf courses with extensive systems to irrigate fairways and greens; and turf farms. Examples of smaller irrigators would include a greenhouse operator with a well and a small farm with a well or pump in a nearby stream. Aquaculture systems are closely related to irrigation and drainage and can be considered in the same way.

Although the focus in irrigation is on rural farming, cities and farms are interconnected. Farms provide food for city dwellers, and often cities provide jobs for farmers. In a city-farm water sharing concept, farmers can agree to idle land in dry years so the cities can have the water. Also, we see the emergence of non-potable water systems in cities, which can provide water for urban farming as well as other non-drinking water uses.

GOVERNANCE OF THE SECTOR

Policy

Water and agriculture should be viewed as a unified system that is managed for greater productivity and to enhance ecosystem services. From an overall point of view, irrigation policy focuses on water-agriculture systems, managing ecosystems, and managing systems for greater benefits (Molden, 2007). Given the economic status of many farmers, irrigation policy is also directed at fighting poverty and improved access to water and farm-to-market systems can improve incomes and increase food security in crop, fishery, livestock, and mixed systems. The water-agriculture system should extend beyond irrigation to rainfed systems as well, and improvements are needed in soil moisture conservation and supplemental irrigation.

Higher value agriculture can focus on integrating crops, livestock, fisheries, and forest management. Management improvements should target state institutions and create methods to deal with tradeoffs and engage stakeholders for a strong collective voice.

These are challenging goals, but they frame the agenda for irrigation water governance, which can be an important coordinating mechanism to link water and agriculture policy. Specific water policy can focus on access and withdrawals, irrigation organizations, and irrigation runoff quality.

The need for food creates an imperative that irrigation water supplies be secured and distributed equitably in spite of competitive pressures such as urbanization and climate change. Given that irrigation water does not normally pay for itself, it is difficult to develop large scale new systems unless they are subsidized. Exceptions might include limited-scale irrigation wells to supply center pivot systems.

Paying for irrigation system infrastructure is challenging under food policy systems.

In the U.S., access to irrigation water is widely available, but cost, legal constraints and environmental pressures present barriers to some. Most western U.S. irrigation water systems are already established and irrigation is viewed as a source of water for reallocation to other users, including the environment. Given the tight regulations on surface supplies, it is unlikely that access will be loosened. Rather than expansion of systems and development of new supplies, the future is likely to see greater use of groundwater and conjunctive use systems, along with city-farm cooperative schemes.

Given the ongoing population growth, there will be opportunities for local farmers to raise vegetables and support urban food markets. Also, new irrigation projects may be considered to respond to demand for golf courses and other recreational venues. Use of reclaimed water for these purposes is increasing.

In developing countries, rural development and alleviation of poverty and social unrest are strongly dependent on equitable access to irrigation supplies. For example, access to irrigation supplies is a critical national issue in Egypt, an arid country that depends on irrigated agriculture on some 8.7 million acres of farm land along the Nile River the Delta to feed a growing population that has passed 80 million and is currently increasing at about 1.7% per year. Water for irrigation comes mostly from the Nile River with supplemental water from reuse of agricultural drainage and use of shallow groundwater. The irrigation system evolved from ancient practices and studies of water management started as early as 1870 in the colonial period, with a 1920 memorandum entitled "Nile control" by the Ministry of Public Works as probably the first documented water policy. Construction of the Aswan High Dam in the 1960s is the defining infrastructure development for water resources management. While the Nile River water has fairly good quality from Aswan to Cairo, the pollution increases as the river flows northward to the Delta region.

Water governance of the irrigation system is based on government control from the highest to the lowest levels. Control of the irrigation system is by the central government through the Ministry of Water Resources and Irrigation (MWRI). The National Water Research Center (NWRC) was established in 1975 to be the research arm of the Ministry. The MWRI has an Irrigation Authority which manages the system, including water allocation to zones and to individual farmers. All aspects of governance from policy through empowerment to control are exercised by a command and control system that begins centrally. So, it is a socialistic system that is not keyed to demand management, so much as it is to central planning.

Water policy in Egypt faces challenges to serve a rising population that lives on a limited land base and has a limited supply of water. MWRI has the governance task to secure and allocate the limited water supply, making sure that adequate water reaches all farmers down to small landholders. The ministry has initiated a National Water Resources Plan to increase water use efficiency, control pollution, and augment water supply. Recommendations from an Egypt Water Use Program led the MWRI to set goals to improve management of irrigation water, minimize seepage losses from canals, manage water table levels, improve the drainage networks, control water better through the distribution system, improve water availability at the tail ends of canals, improve condition of irrigation networks, and increase crop production (FAO, 2007; Egypt Ministry, 2009).

Whereas Egypt has a large scale and centralized system, India has a much larger population and more distributed system based mostly on small landholder situations. There, irrigation institutions have evolved over a long period with state controlled institutional arrangements superimposed over traditional small-scale community irrigation systems.

In a case study located in Kerala State in India, distribution of landholdings shows mostly smaller holdings with about 75% less than one acre. Many landowners are former lessees of large landowners and cropping is mixed with bananas being important. State control of irrigation systems eroded community initiatives and weakened the performance of the systems, but individual and community initiatives have made improvements.

Construction of a government-sponsored water project came after 1947 independence as part of the "Grow More Food" campaign. However, system maintenance was poor and cut water availability to farmers. The people depended on the project but the state departments lacked understanding of the specific crop water needs and farmer needs in different locations. The farmers were not involved in policy making or irrigation management. A Canal Committee was formed, but upstream-downstream cooperation was poor and sometimes involved conflicts as upper diverters took the water before the lower diverters could get theirs.

To compensate for the poor performance of the centralized system, new systems under various ownership regimes have emerged. For example, a Command Area Development Authority has worked to make canal irrigation more efficient and to require Beneficiary Farmers Associations for water allocation and system maintenance. Also, private water sources have been implemented.

The authors concluded based on their field work that success in system management depends on level of participation of water users. The institutional environment requires clear and flexible rules and regulations, enforcement mechanisms, and institutional cooperation (Menon, Paul, and Nair, 2005).

At the highest level where the largest irrigation deliveries take place, regulation of withdrawals from watercourses or well fields is by the means explained in Chapter 4 for overall water allocation. Once the water is diverted into the irrigation delivery system, then allocation is by different authorities who are responsible for fair and equitable delivery to users.

When water flows through a system of canals and ditches, it can be difficult to ensure that supplies reach all the way to the tail end. If the upper irrigators take too much, then there is no water for the lower irrigators and this creates the "tail-ender" problem. This smaller-scale water allocation issue is also explained in Chapter 4.

Irrigation organizations

Access to and management of water for irrigation is facilitated by diverse forms of irrigation companies and authorities around the world. These range from central government control, with lower-level districts, to various kinds of local government districts, cooperatives, and mutual companies. The organizations can be grouped according to the water management and/or farm and social development activities they include. Categories might include integrated management, specialized water management, and multipurpose water management organizations (Sagardoy, 1986).

Integrated management organizations would include state farms, irrigation settlement projects, and irrigation cooperatives. Differences will depend on the degree of involvement of the government and on the wealth of the farmers, who may or may not be able to afford infrastructure.

State Farms may occur where land is nationalized or land reform has taken place. Public irrigation schemes that are mainly controlled by government are numerous and are found in Spain, Turkey, Bolivia, Iraq, Ecuador and Kenya, among others.

Irrigation settlement projects focus on improving the economic and social welfare of landless people or poor farmers by providing them with irrigated land and the means for agricultural production. A cooperative will be established by the volunteer participation of the members and typically, a group of farmers will form an association to develop an irrigation system for their land.

Specialized water management organizations and/or associations are mainly controlled by farmers, and they have varied names (irrigation and water districts in the US, Comunidades de Regantes in Spain, Subak

Irrigation Association in Indonesia, irrigation associations of Afghanistan, the Asociaciones de Canalistas in Chile, and others).

Irrigation associations will be governed by a general assembly of all farmers with a designated Board of Directors. An executive manager will direct work of executing units, which are responsible for functions such as operation, maintenance and administration.

Associations provide a good communication channel between administration and farmers. A good size for them might be some 10,000 ha (2,000–4,000 farmers), according to the study. Governance arrangements will vary by organization type, legal character, water use rights, operational rules, and financial responsibilities.

In irrigation associations farmers participate in decision-making through a democratic process. Recovery of water fees may be better in associations than in public irrigation schemes, and rules and regulations may be respected better. Some associations have juries to enforce rules and regulations.

Irrigation associations offer a path to manage the social issues in sharing water.

Relations between water masters and farmers may be friendlier and more cooperative than when the water masters are seen as government representatives.

In the U.S., irrigation districts show that with good technical and financial resources, associations can be very successful. However, associations have a few problems. The "on-demand" delivery as opposed to "rotation" may create more losses. Old water rights may be an obstacle to efficient water distribution. Transactions among water rights can lead to social unrest. Management and technical skills are probably limited, but empowerment in associations may occur through irrigation extension services.

Public irrigation schemes can be a transitional stage on the way to transferring a project to farm owners. They offer ways to rationalize water distribution and are not burdened by old water rights. Technical management may be improved. In the transitional stage, control may be mixed and fit into a public-private group.

If irrigation organizations are for multiple purposes, they may engage in hydropower generation, navigation, and urban and industrial water supply. Such organizations may be difficult to establish because of various constraints, such as water laws. Farmer associations are based on member contributions and they may not want to branch out. Irrigation associations may have old traditions and not be willing to undertake additional responsibilities.

In the U.S., irrigation organizations range from small cooperatives to giant multipurpose organizations. An example of a medium sized organization is the North Poudre Irrigation Company (2004), a non-profit mutual ditch company in Colorado. In Colorado, the mutual ditch companies are an organized way for water users to cooperate in water delivery and management. Policy for them has been set for decades in Colorado law which authorizes the organization of mutual ditch companies.

The company operates 22 storage reservoirs and approximately 200 miles of ditches. The service area is approximately 42,000 acres, including 28,000 irrigated acres with crops such as corn, sugar beets, soybeans, hay, and feed crops. As of 2004, North Poudre had 646 shareholders, who pay annual assessments. If shareholders do not pay, the company can curtail water deliveries and take actions to recover stock shares. It can enter into financing agreements, such as to use loans from the Colorado Water Conservation Board to improve facilities, with repayment through assessments to stockholders.

Large multipurpose water organizations are common in the U.S. and are discussed in Chapter 10.

The quality of farm runoff and drainage water is an important governance issue, which is mentioned in Chapter 6 under nonpoint source control. Farms discharge large volumes of sediment and nutrients, and smaller quantities of toxic chemicals. Sediment is regularly listed as the number one pollutant in USEPA reports of causes of stream impairment. Runoff from animal production is a source of phosphorous and pathogens in lakes, and agricultural chemicals threaten ground water. Farms can also account for wetland losses and damage to riparian and floodplain environments.

The challenge in regulating farm discharges is how to govern them? In an ideal situation, farms would use best management practices and minimize loss of sediment, fertilizers, and pesticides and most nutrients would be retained in the soil. In the real world, however, farmers are often challenged to manage difficult and variable conditions with financial limitations, and it would be impossible and impractical to regulate them closely.

Quality of drainage and farm runoff is a large environmental issue.

In the U.S. these conditions have been faced by a program that combines incentives and capacity-building with some regulation. The Clean Water Act includes provisions for studies to identify sources and causes of pollution and the types of fertilizers and pesticides that can be used are regulated. Federal and state resource and wildlife agencies cooperate with farmers toward improved management of lands, and the combined effects are supposed to improve water quality. Still, many problems remain unsolved as the stream condition reports show. In addition to sediment, nutrients, pathogens, salinity and pesticides in runoff, the cumulative effects can have large-scale impacts on the environment. For example, there is a large hypoxic zone in the Gulf of Mexico that is caused by Mississippi River discharges that carry farm runoff from fertile lands in the nation's mid-section.

CONCLUSIONS

Irrigation is the dominating water use globally and in regions that are dependent on it. In the U.S., irrigation practices are well-established, and changes are likely to be incremental. As urbanization continues, new subdivisions move into farm regions and urban farming is on the rise. There will be opposition to new large projects, but reallocation of irrigation water and increased uses for urban farming and recreational uses is evident.

In many developing countries, particularly in the 20–40 window, irrigation is an important farm input and is connected to the social fabric of local societies in the same way as banking, schools, and other core services. Irrigation may have been practiced for millennia, but population growth and the need for increased food security may continue to drive new irrigation projects.

Whereas governance of water and wastewater organizations is facilitated by their identification as utility enterprises, irrigation organizations are more diverse and more like mutual organizations among users. Even the large organizations exist to serve networks of politically-connected users.

Ideally, irrigation user organizations could govern themselves and comply with environmental regulations but the incentive structure is to take all water available and not to pass along essential supplies to downstream users. This problem is normally handled in the same water as that for other withdrawals.

The economics of irrigation water governance are problematic. Farmers are often not able to pay the cost of irrigation water and in many cases farm prices are not high enough anyway. Subsidies in food policy can work against good water management. If policies can serve to raise farm income, they can promote the reallocation of water toward higher value uses.

The empowerment feature of water governance is important in irrigation because farmers often lack the means to implement their own projects and rely on government authorities for assistance. Governance roles to facilitate access to irrigation water focus on: including agriculture, where appropriate, in planning and development projects; ensuring that agriculture's interests are represented fairly in forums and planning programs; and facilitating discussions about the possibility of water-sharing programs, such as city-farm drought contingency plans.

Irrigation water quality issues will continue to be difficult to address. Programs of empowerment, subsidy, and light regulation seem to be the best mix.

Environmental activists are critical or irrigation water use, but governance should assure that urban public opinion is not misled about the value of water applied to irrigation. For example, in the U.S. it is widely believed that agriculture has too much water, is wasteful and polluting, and even receives subsidies for its irrigation projects. The issues are much more complex than this, but bias against agriculture can be formed as a result of these images.

Where rural incomes and food production are critical issues of national security, governments must take actions to improve access to irrigation water. Solutions to rural hunger and poverty demand better access to irrigation water for farmers with small landholdings.

Irrigation requires many reforms. Jan van Shilfgaard (1992) summed it up in a lecture at the U.S. National Academy: "In the future, irrigation worldwide will gain further in importance, and better water management is essential for human survival." We can learn from our errors. There are clear limits to the insults the environment can absorb…" "We should face up to the fact that equity has suffered, that third parties often have not been served well, and resolve to do better." "As water professionals, we should take the initiative towards strengthening the irrigation industry in a manner that is equitable to all parties, compatible with greater social goals, in deliberate balance with environmental needs and sustainable for generations to come."

Postscript

As a child, I spent a lot of time on farms but they were rain-fed and did not require irrigation. The main crops included cotton, corn, soybeans, and many varieties of vegetables. My introduction to irrigation came with our move to Colorado in 1965.

Colorado is a dry state and crops will not survive there without irrigation to supplement the some 14-inches of rain and snow water that occur annually. The exception is a form of risky "dryland" farming, where farmers leave fields fallow to build up soil moisture, and then grow crops by combining soil moisture depletion with annual precipitation.

After my initial introduction to Colorado water management—which mostly evolved from a system to supply water for farms—I came to see how integral irrigated farming is to societies where it is practiced. In Colorado, it made the difference between an area that seemed like a desert to the early explorers and the settled area we have now. Other parts of the world have similar experiences, but some irrigated societies go back many centuries earlier than Colorado.

I was lucky to see Colorado irrigation up close through a number of engineering experiences. These included relocating irrigation ditches for highway construction, avoiding using irrigation systems for stormwater conveyance, working with ditch companies to convert irrigation water rights to municipal uses and doing stream-aquifer studies to implement a new Colorado groundwater law.

In 1969 I went with a group of students and teachers to Chihuahua, Mexico and saw how this dry area was totally dependent on irrigation. This whetted my appetite to learn more about how irrigation could be used beneficially for the economic and social development of water-short areas. I noticed that this was a large issue for the programs at Colorado State University, which had developed irrigation engineering programs to support pioneer farmers in Colorado. By the time I became involved, the university's programs were being extended to other nations through foreign assistance programs, and my professors told fascinating stories about studying the big Link Canals in Pakistan, as well as large systems in the U.S. In 1975 I had the opportunity to participate in a water planning mission in Egypt where we studied irrigation along the Nile River and in the Delta. Later, I worked in the same department as Colorado State professors and Egyptian scientists who were evaluating many water problems faced in the irrigation systems of Egypt and Pakistan. I supervised the thesis research for graduate students who worked on irrigation problems in the U.S. and other countries. As examples, one of these reviewed the irrigation systems of Indonesia (Sjarief, 1995) and another considered the economic performance of irrigation systems (Rijsberman, 1987).

At long last I got the chance to do some irrigating myself when we purchased a small alfalfa field, which as a water right under a local mutual ditch company. This gave me an opportunity to jump through the hoops of scheduling the water, cleaning the ditch, operating the turnouts, and seeing how we could irrigate the whole field with a limited supply of water. Our flood irrigation system does not work very well and I have ideas for improvement, but they are not economically feasible with the alfalfa crop. So, at least I am seeing up close some of the problems that farmers face, but my experiences pale when compared to the real problems of managing irrigated agriculture, of course.

I've gained an appreciation for irrigated farming and marvel at how little most people understand about it. It's common for new arrivals in Colorado to look around at irrigation systems and quickly conclude that the environmentalists are right: irrigation is a waste of water. This seems so short-sighted to me, when the real issue is much more complex. It's true that the economic return from the application of a unit of irrigation water is normally less than in municipal and industrial uses, but this ignores many social and environmental issues, as well as how river and watershed systems work. The lesson is that those who govern irrigation systems require careful study before jumping to conclusions. The impacts are great and much is at stake around the world.

REFERENCES

Buyalski C., Ehler D., Falvey H., Rogers D. and Serfozo E. (1991). *Canal Systems Automation Manual*, Vol. 1. U.S. Bureau of Reclamation Water Resources Technical Publication, Denver.

Egypt Ministry of Water Resources and Irrigation (2009). *Minister's Message*. http://www.mwri.gov.eg (accessed 26 December 2009).

FAO (2007). *Egypt's Experience in Irrigation and Drainage Research Uptake*. International Programme for Technology and Research in Irrigation and Drainage.

Menon V., Paul A. and Nair K. N. (2005). *Dynamics of Irrigation Institutions: Case Study of Village Panchayat in Kerala*. http://www.cds.edu (accessed 24 December 2009).

Molden D. (ed.) (2007). *Water for Food, Water for Life. A Comprehensive Assessment of Water Management in Agriculture*. International Water Management Institute, Sri Lanka/Earthscan Publications, London.

North D. C. (1990). *Institutions, Institutional Change, and Economic Performance*. Cambridge University Press, Cambridge.

North Poudre Irrigation Company (2004). *Feasibility of Rehabilitation of North Poudre Reservoir No. 1*. http://cwcb.state.co.us (accessed 24 December 2009).

Rijsberman F. R. (1987). *Development of Strategies for Planning for Improvement of Irrigation System Performance*. Ph.D. Dissertation, Colorado State University, Fort Collins, Colorado.

Sagardoy J. A. (1986). *Organization, Operation and Maintenance of Irrigation Schemes*. http://www.fao.org/docrep/x5647e/x5647e06.htm (accessed 26 December 2009).

Sjarief R. (1995). *Holistic Monitoring & Evaluation of Irrigation Development: A Model for Management Control*. Ph.D. Dissertation, Colorado State University, Fort Collins, Colorado.

US Geological Survey (2009). *Water Use in the United States*. http://water.usgs.gov/watuse/ (accessed 16 December 2009).

van Shilfgaarde J. (1994). Irrigation: a blessing or a curse? *Agric. Water Manage.*, **25**(3), 203–219.

Chapter 8

Instream flows and dam operation

Within the world's water issues, a silent crisis is occurring in the quantity and quality of stream flows. Of the many millions of cubic kilometers of earth's water, only a tiny fraction is in streams, lakes and estuaries. But this tiny fraction serves as the nursery for myriad species to stock the food chain and is therefore a critical part of our survival system. Why should humans spoil such a valuable and essential support system?

This chapter provides insights into why this occurs and what must be done to prevent such despoliation. The chapter is about instream flows, but this really means the management of shared water resources as they move along the hydrologic cycle. Many issues are embedded in the concept, and they cry out for cooperative solutions. Given the nature of human incentives it will be difficult to bring the solutions about and they present a formidable challenge to water governance.

Whereas water and wastewater involve clearly-defined services, the governance of flows in streams and rivers is more ambiguous. The interdependent uses of water along a stream require many decisions about withdrawals, discharges, and dam releases, while considering the attributes of the natural flows.

We call these "instream flows" to create a focal point for identification of this water management problem. In many ways, their management is the most stringent test of water governance and of integration in water management. This occurs because the quantity, quality, and timing of instream flows depend on both success of programs such as control of withdrawals and discharges and on coordination among many stakeholders along the streams.

This chapter explains the complexities of instream flow management, and offers suggestions for governance of this sector, with emphasis on water quantity. Governance of environmental water quality is explained in Chapter 6.

INSTREAM FLOW SETTING

Streams and rivers can be considered as water "commons," and they require coordination because no single authority is normally responsible for them. These water commons are the organizing features of watersheds, which are the natural and logical units for water management. The logic of managing water within watersheds has been recognized for many years, but the difficulty inherent in water crossing political boundaries is always a challenge.

Figure 8.1 Comprehensive watershed diagram

The drawing of a watershed and stream network shown in Figure 8.1 is from a U.S. water policy study from 1950, and it illustrates the many activities that take place within a watershed (President's Water Resources Policy Commission, 1950). To focus on instream flows, look at the main stream, which begins with glacier melt into a multiple purpose reservoir and continues to become a larger stream that supports navigation and other uses. The smaller stream to the right, which is experiencing a rainstorm, drains a separate watershed. For it, you can imagine that upstream irrigators are taking all the water and leaving none for downstream users or the environment. For the larger stream, you can imagine that the dam operator is not releasing enough water for navigation or any other downstream uses. In both of these cases, instream flows would not be managed adequately, thus causing many problems.

From the watershed diagram it is a short step to create a line drawing such as Figure 4.1, which shows many diverters and dischargers. These, along with dam releases and natural hydrologic phenomena, determine the quantity and quality of instream flows at any location and point in time. You could go from this drawing to the next step of an analytical expression of the watershed in the form of a mathematical model of the flows and water qualities.

The instream flows provide several critical services for water users and the environment. They provide "carry water," which can be considered like a vehicle to transport water to points of diversion. To visualize this, imagine a dry streambed and a water user who needs to divert a water right to a farm. If the water had to be transported from an upstream reservoir to the point of diversion, a lot would be wasted in losses in the dry streambed. Now, if many such diverters pool their flows the losses will be about the same, but they will be distributed over the many users. Therefore, the combined instream flows are providing a service to all the users.

These instream flows are also providing a service to wastewater dischargers because the larger flow rates provide dilution water for the discharges from treatment plants and non-point runoff.

They also provide a service to the environment, which enjoys the benefits of greater flows with higher quality. This can be called an "ecosystem service" (Molden, 2007).

Other benefits of instream flows accrue to hydropower generators, recreational facilities, and navigation interests. All of these users depend on the water as it flows through the "water commons."

It is evident that providing instream flows for these diverse water services and needs creates an imperative for what Gilbert White (1969) called the management of water for "multiple purposes" by "multiple means."

THE INSTREAM FLOW SECTOR

To be specific, the instream flow sector includes water uses for conveyance, environmental water quality, water for habitat, hydropower, navigation, and recreation, including aesthetic purposes. The participants in the sector are the water managers who depend on and control instream flows and the planners, regulators and coordinators who influence the flows. Table 8.1 outlines their roles and responsibilities.

TRAGEDY OF THE COMMONS

It is not practical for governance to strive for such an overall authority, so instream flow governance must work indirectly through the actions of its multiple participants. In many ways, it is a poster child for the concept of "shared governance" (see Chapter 16).

It is evident from the list that no one is in overall charge of instream flows and this creates the behavioral problem called the "Tragedy of the Commons" as it relates to flow in streams. This phrase was coined in a 1968 article in the Journal Science to explain why, when people look after their own interests but neglect the public interest, the commons suffer (Hardin, 1968). His example was about animals grazing in the town commons, which are quickly depleted, but it can extend to many other examples of public resource issues. In the U.S. a historical example is the need to ration the use of open range lands for cattle grazing. In the case of instream flows, it is in the self-interest of diverters to take as much water as they can and they have little incentive to conserve so that adequate flows remain in the stream for all to use.

Table 8.1 The roles and responsibilities of water managers

Participants	Roles and regulators
Water conveyance managers	Their role is to assure "carry" water to the points of use. They include water diverters, dam owners, and dischargers.
Wastewater dischargers	They provide water suitable for dilution and add to instream flows.
Environmental and fish and wildlife agencies	A diverse set of agencies focuses on water for plants, fisheries and animals and influence instream flows for these objectives.
Hydropower generators	Power agencies control instream water to generate hydroelectricity through dams and diversions.
Navigation agencies and companies	Navigation entities require minimum depths for commercial shipping and transport.
Recreation agencies and companies	Recreation managers may control streams and lakes for public swimming, boating, fishing, and aesthetic purposes.
Regulators	State instream flow rules (where they exist) are the most direct form of regulation. Other forms of regulation include: control on withdrawals and discharges, federal environmental controls, dam safety and operations, FERC relicensing, PUCs, water transportation rules, and health and safety.

GOVERNANCE

Policy

As governance for instream flows involves diverse interests, policy should focus on assuring that effective coordination mechanisms are in place to achieve satisfactory outcomes for all. Some elements of the process may be handled by coordination among water diverters, dam owners, and dischargers, so suitable forums for the coordination to occur should be provided.

This provision of forums to resolve issues that fall between the cracks of other governance programs, together with additional measures to deal with issues such as low flows and TMDL problems, comprise the main approach to instream flow governance.

Low flow policies are particularly relevant for instream flows because conditions are more critical during drought periods. It is important to remember that drought can be induced by natural or man-made causes, so policy should be adequate to respond to the natural causes and to avoid the man-made droughts.

Agencies with authority over wastewater dischargers, hydroelectric dams, and environmental instream flows have influence on the overall stream flow regimes. Although these agencies may be independent of each other, they can coordinate their policies and rules.

Empowerment

Empowerment for the instream flow sector should begin by defining its objectives and identifying roles of the diverse participants. Given that its outcomes depend on actions of water managers who are pursuing other interests, the Tragedy of the Commons will occur unless they cooperate in achieving the common goal of healthy instream flows.

A productive route to empowerment for the sector might focus on technical and financial assistance to the entities with important roles, including the regulators. The assistance can be aimed at defining objectives, identifying roles, and providing tools and incentives for positive behavior.

A principal question would be: who has responsibility for empowerment in this sector? This question underscores the importance of identifying the issue and explaining it to leaders and policy makers. Unless the need for instream flow management is understood, no action will occur and stream health might revert to the lowest common denominator.

State water agencies, river basin and regional authorities, and watershed groups are logical entities to take the lead on empowerment. Water users can also take it on themselves to implement mutual aid and to cooperate on instream flows.

Controls

Controlling includes planning and coordination as well as regulatory restrictions. Agencies can use a range of tools to influence instream flows.

When an authority issues a water use permit, it may designate minimum instream flows, both for normal and drought conditions.

There is considerable public interest in having wise policy on instream flow maintenance. Regulators need legally–binding ways to restrict water withdrawals to leave water for future water users, for water quality and habitat, and to accommodate needs of instream users, such as navigators and hydropower generators. If f streamflows are too low, new industries and wastewater dischargers cannot get permits, fish and wildlife may suffer, and a present water user may not exercise water conservation, thus taking water that could be used by a user who comes later, creating jobs and economic benefits.

CONCLUSION

Instream flow management fits the logic of managing water within watersheds, which has been recognized for many years. It addresses the water commons in stream networks, from headwaters to tailwaters and even to the oceans. It is concerned with both the water quantity and the water quality in streams, but they are dealt by separate governance programs.

The management of instream flows is a stringent test of water governance because no single authority can control them and water users have the incentive to use the resources to extinction without paying. both on the water allocation and water quality sides.

Instream flow management has many participants and requires shared governance, which is a challenge (see Chapter 16). The participants include the water conveyance managers, the water diverters, dam owners, and dischargers, as well as instream water users and regulators.

In the U.S., the need for instream flow management is recognized, but the management programs require a lot of work. Instream flow programs are hard to administer directly because the flow is not controlled directly by some permit or regulation but is the result of all of the withdrawals, returns, and natural inputs. It requires a coordinated approach more than it does a command and control approach. The requirements for a solution include assessment of the instream needs, control of allocations, low flow programs, water quality programs, and coordination.

Empowerment for instream flows should begin by convincing people that it is important. In particular, environmental agencies have a strong vested interest to push for improvements. For example, in the U.S. and Canada, as group known as the Instream Flow Council (2010) represents state and provincial fish and wildlife agencies to promote instream flow programs. The council focuses on the technical, institutional, legal, and public involvement activities of fish and wildlife agencies to promote healthy streams.

It is difficult to estimate ecological needs, and people do not like to leave water for fish when they perceive they need it for economic purposes. There is therefore a need for modeling and study, which are essential to provide the information base to support the needed coordination. This is similar to the need for modeling in the TMDL program, which must be supported by extensive studies to avoid harmful decisions.

Postscript

Probably my first awareness of instream flow issues was the despoliation of a small creek in Alabama by a gravel mining operation. Later, after returning from Army service, I discovered my favorite river dammed up into a lake to create a government-sponsored navigation pool. Later, I saw the western instream flow problems when, after moving to Colorado in the 1960s, I noticed a diversion that completely dried up the Colorado River in a section of the beautiful Glenwood Canyon. My reaction was "What could have happened to cause this calamity?"

The cause was a diversion by an electric utility that had the legal right to take 100% of the stream into a long pipeline to a hydroelectric plant. It was a diversion with only economic purposes but with severe environmental consequences. That was a result of the Colorado system of water law (see Chapter 14). Take heart, some wise people have not worked out a compromise and some water trading to leave water in the stream and reverse that environmental disaster.

A few years after that, I was working in the North Carolina state government and had authority to regulate some aspects of stream diversions and discharges. I noticed that there was a community of people with the same interest as mine in a total approach to managing streams for all purposes but we had to fight cities and water developers who wanted to skim off all of the water and to pay the absolute minimum.

As I am writing this (in 2010) we have an instream battle going on in Colorado where a water district is proposing to skim flood flows to create a reservoir to serve water needs of cities and districts. They have a good plan but it will diminish some flows in our local small river and a series of agreements is needed to move toward the total approach to water management. The environmentalists are fighting the proposal tooth and nail, but they do not have a realistic plan for improvement and only seem to oppose water development. The Corps of Engineers has been engaged to coordinate shared vision planning here, but the likely outcome is unclear. Meanwhile, the water district is paying millions of dollars for an Environmental Impact Study, which may result in some kind of limited permit with stringent conditions and lead to a protracted legal battle.

It seems to me that the governance system for instream flows should avoid these conundrums, but it is held down by institutional constraints and perverse incentives related to the Tragedy of the Commons. There seems to be no choice but to continue to seek incremental improvements as we deal with instream flow issues one at a time.

As a final note, management of instream flows as a water commons has similarities to the challenge of managing air quality. The "cap and trade" regulatory approach has shown merit for allocating the right to emit sulfur dioxide and might be used for carbon dioxide in the future. The rights to divert, return and pollute water could eventually become a market device for managing the water commons. This seems a long way off, but seems to be a possibility worth discussing.

REFERENCES

Hardin G. (1968). The tragedy of the commons. *Science*, **162**(3859), 1243–1248.

Instream Flow Council (2010). *About Us*. http://www.instreamflowcouncil.org/index.htm (accessed 19 June 2010).

Molden D. (ed.) (2007). *Water for Food, Water for Life. A Comprehensive Assessment of Water Management in Agriculture.* International Water Management Institute, Sri Lanka/Earthscan Publications, London.

President's Water Resources Policy Commission (1950). *A Water Policy for the American People.* U.S. Government Printing Office, Washington, DC.

White G. F. (1969). *Strategies of American Water Management*. University of Michigan Press, Ann Arbor.

Chapter 9

Flood and stormwater management

Flood and stormwater management are part of the water sector, but they are distinct from the water supply and wastewater management sectors in the sense that the service they provide is to handle excess water, rather than to supply water. Other than this distinction, they are similar to the other water sectors in that they involve similar water agencies and infrastructure. They are often bundled with the other water services, and stormwater is closely coordinated with urban planning and development. Flood control management is often bundled with navigation and the operation of dams for hydroelectricity.

Governing flood and stormwater services is more difficult than for utilities because responsibilities are normally more diffused. For example, in the 2004 Hurricane Katrina disaster in New Orleans, responsibility to maintain the levees that failed and caused the disaster was dispersed over different local governments, as well as the US Army Corps of Engineers. Stormwater issues are often inter-jurisdictional, with land uses upstream causing damages downstream.

The Hurricane Katrina damage was caused by a failure in governance as much as it was by failed technical systems.

Flooding causes much damage and misery and requires high priority attention from officials. Often, improvements can be made without great cost, but they require effective management and governance.

THE SCALE OF FLOOD AND STORMWATER PROBLEMS

Globally, flood problems are very serious and the deaths, disease, damage and misery caused by flooding are of awesome scale. In local areas, stormwater problems displace people, disrupt transportation, damage property and generally degrade urban life unless adequate drainage systems are provided.

Flood control has general purposes and benefits that extend well beyond individual households. It is normally impossible to protect one home from floods without protecting nearby structures, although single structures can be floodproofed. However, stormwater management is more local in nature, and its benefits can sometimes be assigned to individual properties.

Basinwide and local solutions are needed for flood and stormwater problems. While flood disaster mitigation requires locally-focused planning, upstream watershed changes can exacerbate flooding and require controls to protect people living downstream.

The chapter explains how the flood and stormwater sectors have similar characteristics but require different management and governance approaches. It shows the distinctions between the services and summarizes the main governance requirements for each one.

THE FLOOD AND STORMWATER SECTOR

The need for this sector's services occurs because rainwater must be managed to avoid damages to people and property and to enable society to function without unreasonable disruptions to transportation and other systems. Hence, the focus of the sector is on the management of excess water. The problems it deals with range from great floods along major rivers to relatively minor drainage problems within communities, and this explains the distinction between the two parts of river flooding and stormwater management.

River flooding and stormwater drainage are problems of a different scale.

This distinction is shown in Figure 9.1, which illustrates stream flooding, urban stream flooding, and stormwater collection. These different categories prove to be important in establishing governance systems because they operate under different authorities and rules. For example, in the U.S., stream flooding in a rural area might be more the concern of federal authorities, while urban stream flooding and stormwater collection will be under local jurisdictions, often several inter-related jurisdictions. Control of stormwater quality has also become a major issue, and it is a concern of all three levels of government.

Whereas both flood control and stormwater management deal with excess water and both are protective in nature, they operate at different scales. Both are linked to environmental and recreational purposes, and stormwater also involves water quality control and links to urban amenities, such as ponds and open space.

Flood control policy in the U.S. has evolved over about a century. In his book on Strategies of American Water Management, Gilbert White (1969), considered a father of U.S. flood policy, explained that flood control involves "multiple means," whereas other water sectors normally are involved in "multiple purposes."

Given its large scale, river flood protection and flood plain management in the U.S. draw in federal organizations such as the Army Corps of Engineers, with its flood control reservoirs, the National Weather Service, with its flood forecasting and warning system, and the Federal Emergency Management Agency, which handles flood preparedness. State and local governments also have flood-related organizations, ranging from state flood coordinators to local offices for flood preparedness, response, and emergency management.

Stormwater management is mostly a local concern and its major organizations tend to be local agencies. Stormwater systems comprise minor, smaller systems of pipes and other conveyances and storage sites, mainly within cities but also used in any place where drainage of sites and facilities is needed. The main program of national interest is the control of the quality of urban stormwater runoff, which is explained in Chapter 6.

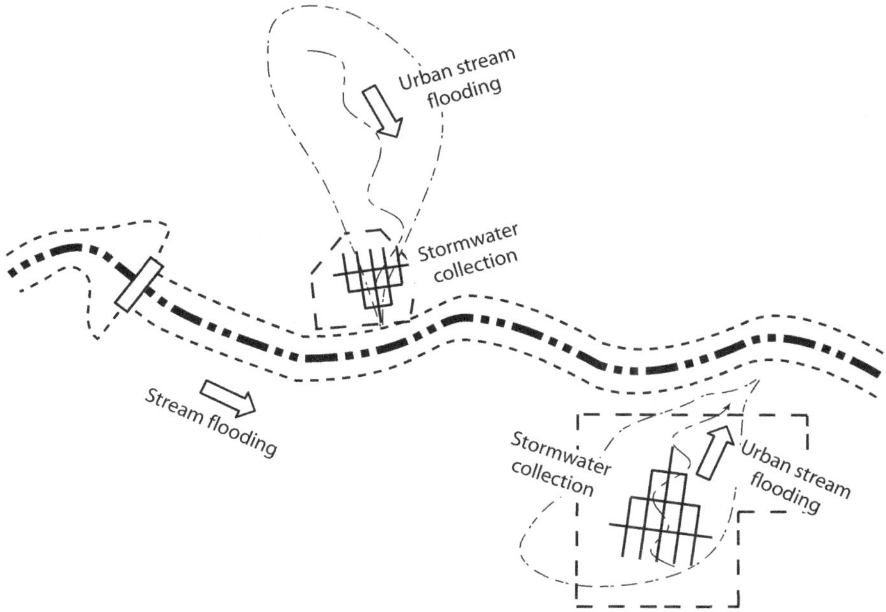

Figure 9.1 Flood and stormwater systems

GOVERNANCE OF FLOOD AND STORMWATER SERVICES

Flood and stormwater policy

National flood policy in the U.S. continues to evolve. Early policies were set by programs of federal agencies, mainly the Corps of Engineers. The 1968 Flood Insurance Act launched a new, federally-subsidized insurance program to protect properties at risk. After the devastating 1993 Mississippi River flooding, a national study was launched and recommended more emphasis on non-structural and environmentally-friendly strategies (Task Force on the Natural and Beneficial Functions of the Floodplain, 2002). The New Orleans disaster from Hurricane Katrina also stimulated new policy studies, especially with a focus on safeguarding levee systems.

Stormwater policy emanates from housing, transportation, and environmental policies. The goals are to use stormwater systems to enhance living environments and to provide for minor flood control so roads and structures are free from damaging drainage flooding and to prevent the discharge of pollutants.

In the U.S., policy for levels of stormwater service is mostly of local origin, and one of the most influential policy drivers was the development of the Urban Drainage Criteria Manual, which was prepared by a local district, the Urban Drainage and Flood Control District in Denver, Colorado (see Chapter 10). Also, stormwater policy evolves through housing and subdivision regulations, and much of today's emphasis on sustainable cities deals with stormwater. Debo and Reese (2003) have presented a comprehensive text about stormwater regulations and policies.

Empowerment

Capacity-building for stormwater and flood management involves institutional, organizational, and human resources development. Given that the programs are not of the highest priority among water agencies, they are sometimes not well-organized. In developing countries, where flooding and stormwater problems can cause great suffering and losses, capacity-building is of special importance (Katsuhama, 2010).

It is natural that flood agencies should be at levels of government high enough to have authority over large stretches of rivers or basins. Thus, in the U.S., the Corps of Engineers, the Federal Emergency Management Agency and the National Weather Service have important roles at the federal level. Flood control cannot be established as a self-supporting enterprise, and it needs a lot of coordination and emphasis on emergency preparedness.

Local institutions for stormwater management are often city and county governments, which also have responsibilities for public works functions such as streets. These agencies often need assistance and training for stormwater technologies and management.

Control

Flood control and stormwater involve regulatory controls, but they tend to be land use controls rather than direct regulations about excess water. In the case of flooding, the controls are of land use in flood plains and along coastal areas. The flood insurance program is both a regulatory and an assistance program. Dam safety is also a flood-related regulatory program. In some ways, interjurisdictional coordination and planning and projects comprises a control mechanism.

In the case of stormwater, there are many regulations about development of urban drainage systems, impervious surfaces, and upstream-downstream liability issues through drainage law. Hydrologic alterations create are some of the most significant insidious environmental effects from land development and some of them are controlled through the 404 program, under the CWA.

CASE STUDIES

U.S. Flood Insurance program

Flood policy in the U.S. has evolved from the structural era, which focused on building flood control dams as the nation developed. After flooding in the decades after 1900, Congress authorized flood control works in the Flood Control Acts of 1917, 1928, 1936, and 1938. The emphasis in these acts was on works such as dams, levees, and channel modifications. By the 1960's the federal government had completed many projects, especially through the Corps of Engineers.

The nation was aware of the shortcomings of flood control structures, and as a result of Gilbert White's work and other developments, by the 1960's the emphasis had changed to non-structural measures such as zoning, flood forecasting, and flood insurance. White's (1942) dissertation at the University of Chicago in the 1940s was entitled "Human Adjustment to Floods," a descriptive title that presaged today's focus on non-structural solutions.

Gilbert White's foresight led to advances in non-structural flood control.

The effect that White's dissertation had on flood policy was evaluated in a 1997 paper, which explained "Whatever its exact statistical contribution, the USA and the world are safer because it (White's dissertation) was written, and because the author continues tirelessly to advocate, personally and through his warm encouragement of others, the perfectibility of human settlements on the earth" (Platt, 1997).

In 1960, the Corps of Engineers began to produce floodplain information reports as a service to local governments. In 1966, a Unified National Program for Managing Flood Losses by prepared by the Johnson Administration, and it made recommendations for federal agency actions such as research and information. In 1968, the Flood Insurance Act was passed, and it became the cornerstone of flood mitigation efforts.

In 1976, a report from the US Water Resources Council cited coordination as the weakest component of current management efforts in flood control. In 1979, a federal Inter-Agency Floodplain Management Task Force expanded the strategies to include restoration and preservation of natural values. In 1986, the Federal Emergency Management Agency (FEMA) assumed responsibility for the Task Force and prepared a federal disaster mitigation strategy with attention to coastal zone flooding.

The Federal Emergency Management Agency coordinates flood preparedness strategy in the U.S.

Fragmentation of flood authority in the federal government continues. A number of agencies participate in flood insurance studies and other flood-related activities.

HAZUS development

With support from FEMA, the National Institute of Building Sciences (NIBS) prepared a suite of software programs named HAZUS for damage loss estimation from earthquakes, floods, and wind. It has its own user groups and website (Federal Emergency Management Agency, 2010a).

The current version of the software is embodied in HAZUS-MH (Hazards U.S. Multi-Hazard), which is intended to be a standard methodology to estimate potential losses from earthquakes, hurricane winds, and floods across the nation. HAZUS-MH uses GIS software to map and display hazard data and loss estimates for infrastructure. The intended use is for decision-making for mitigation plans and policies, emergency preparedness, and response and recovery planning.

I participated on the advisory group for HAZUS-Flood, which was developed by a team of consultants working under NIBS. The software development process was complex, but it resulted in a workable tool for use by flood planners and provides technical assistance to local and state governments in their mitigation planning.

Dam safety

The national dam safety program is a good example of a combination of regulation and empowerment. While dams can fail under different conditions, they are of special concern during flooding. This creates a need for a dam safety program to oversee risk of dam failures and to include issues such as spillway adequacy.

While dams have been in existence for many centuries, early builders did not recognize the need for scientific approaches to spillway design. During the 19th and 20th centuries, scientific hydrology developed and risk assessment of dams began to be possible. In the 1930s the approach to finding an upper limit to precipitation emerged, and there was considerable discussion about using it for dam studies, but no agreement emerged until the National Dam Inspection Act of 1972, which charged the Corps of Engineers with developing a system of dam and hazard classifications and spillway design floods. Although the system was developed as a guideline, it was adopted by many states. In the 1980s risk analysis emerged as a different approach to the spillway design flood question. It seeks to minimize the total project cost, including the expected cost of damages due to failure, but it has not been widely embraced (Dubler and Grigg, 1996).

After the Water Resources and Development Act of 1996, the nation began to work through the National Dam Safety Program (NDSP), which is administered by FEMA (2010) and is a partnership of the states, federal agencies, and other stakeholders. In addition to dam inspections, it provides for grant assistance for improvement of state dam safety programs to regulate most of the 79,500 dams in the U.S., dam safety research and training. The Dam Safety and Security Act of 2002 reauthorized the NDSP and added provisions to safeguard dams against terrorist attacks.

Flood mitigation in Indonesia: a systems approach

Solving a complex problem such as in the Jakarta, Indonesia region shows required elements of governance and management in the flood control sector of a developing country. Jakarta has daunting flood control challenges, which have been studied by several Ph.D. students at Colorado State University (Hadimoeljono, 1992; Rahardjo, 2002; Akmalah, 2009; Katsuhama, 2010).

These studies showed that flooding and stormwater problems are caused by both natural events and human activities. In particular, urbanization, lack of infrastructure, inadequate operation and maintenance of drainage systems, and related social factors, such as poor solid waste management are blamed.

With its high rates of growth, municipalities in Jakarta have not been able to cope and are unable to provide for basic needs. Jakarta is vulnerable to flooding from both climatic factors and due to its location on alluvial lowlands. While the city has taken some actions and invested in infrastructure, disaster preparedness has mostly focused on technical solutions. Like some other large cities in humid and low-lying areas, the city suffers frequent and heavy flood damage.

Governance factors show that while flood protection policy is in place at the national and regional levels, different approaches are needed. The linkage of the flood problem to social and governmental conditions means that urban, social, and economic policies are as important as flood policy itself.

Of course, a legal framework to provide direction is needed, along with effective implementation, compliance and enforcement. However, capacity-building is the main need, with a focus on institutional capacity, financial capacity, and workforce capacity. Institutional capacity requires improved functioning of flood preparedness and response authorities and attention to community cohesion and preparedness. The social element is challenging as stakeholders see things through different lenses. Coordination and cooperation among stakeholders is needed to support community participation to prevent and mitigate the risk.

The financial capacity of the government to mitigate flood hazards is not adequate to solve the problem with structural measures alone and an integrated risk-based approach is required to reduce vulnerability over time with a mix of structural and non-structural measures. Given the lack of financial resources, coupled with the urgent need to improve the problems, the main solution should not be financial anyway, but it should focus on fixing community issues that exacerbate the flooding. Workforce capacity requires that officials with responsibility in the flood sector are trained to deal with a wide range of issues.

The control function focuses on regulation of land use and on planning and coordination. These are difficult due to the large numbers of migrants who occupy settlements along the waterways and who are difficult to displace.

Flood problems such as those in Jakarta are experienced in other large cities in developing countries. Katsuhama (2010) explained the elements of institutional and technical capacity that are needed to address them through governance programs. Clearly, regulation is a part of the solution but much more is needed through empowerment.

Interjurisdictional stormwater program

In Chapter 10, Denver's Urban Drainage and Flood Control District was described as a multijurisdictional district that plans major drainageways and floodplains, funds and manages capital improvements, operates a maintenance program, and assists in water quality planning. In performing these functions, it is involved in several elements of governance.

In the first place, it was a governance gap that led to the District in the first place. Back in the late 1960s the region's leaders realized that Denver had not responded well to the large 1965 flooding events and one of the biggest problems was lack of coordination among local governments. Therefore, the UDFCD was envisioned as a multi-jurisdictional entity to provide the needed coordination. In its enabling legislation, the Colorado General Assembly spelled out policy for the District's programs.

Early on, the District was involved in enabling the local flood control programs in that it created design standards through a widely-accepted Urban Storm Drainage Criteria Manual, which is still in use. Through its program of financing planning studies, it helped to train many local officials and consulting engineers in waterway master planning. Its mill levy provided a small but valuable source of financial assistance for the local governments it served.

The District is not a regulatory agency but its programs of coordination provide a de facto regulatory mechanism that has equalized drainage master planning and stormwater management in the Denver metro region.

CONCLUSIONS

The chapter began by explaining how the flood and stormwater sectors are different from the other water management sectors because they handle excess water, but they involve many of the same water agencies and infrastructure. Flooding and stormwater are serious problems in most countries, and in the U.S., the damage caused in New Orleans by Hurricane Katrina showed big gaps in preparedness and generated a lot of soul searching. In developing countries, even minor stormwater events can sometimes create miserable conditions and lead to public health problems and hindrance to economic development.

As water issues, flooding and stormwater touch several other sectors, including emergency management, land use, and transportation, in particular. Policy must be multi-sectoral, as shown in the U.S. approach to flood management. This policy is inherently difficult to administer and gaps such as were evident in the Katrina disaster can result. Governance cannot involve simply regulation but requires much empowerment, in particular the development of institutional capacity.

Postscript

My first personal experience with flooding was about 1955, when heavy rains backed up the Alabama River and covered our roads. I knew about local stormwater flooding, as you could see it frequently, but this was my first large scale river flood. However, I didn't realize how much a flood was going to affect my future life.

This was the June 1965 Colorado flooding that caused great damage in the Denver area. It created engineering work for me, and we moved to Colorado where I have worked mostly for the last 40 plus years. Although Colorado is a dry state, flooding is a big issue. I noticed that the 1965 flood gave impetus to stormwater planning and led to establishment of the Urban Drainage and Flood Control District and its influential Urban Storm Drainage Criteria Manual (see Chapter 10). My engineering work brought me into contact with leaders in stormwater and flood management, and I soon began to learn a lot more about advances in the field.

One of those advances was passage of the 1968 Flood Insurance Act, which I studied carefully to learn about flood plain regulation (see Chapter 14). I did several floodplain studies with an early version of what later became the standard U.S. computer model for that purpose. Stormwater management started to become "green," as planners sought to replace concrete channels with open space. I noticed that stormwater and flooding were complex to govern, and I found opportunities to deliver short courses about them.

During the 1980s I was appointed to our county flood plain regulation board, and I helped the administrators to make decisions about which properties to zone into the flood plain. A decade later I was asked to join a national review committee to help create the HAZUS-flood software that was mentioned earlier in the chapter. This was a great opportunity to work more closely with Gilbert White, who was also on the committee and who never failed to remind us of the beneficial effects of floods. A memorial to Gilbert was dedicated in 2010 along Boulder Creek in Boulder, Colorado.

Over the years I noticed that flooding got worse in spite of our efforts to control it. The greatest disasters of the last few years were the 2004 Tsunami in South Asia and Hurricane Katrina in 2005, which devastated New Orleans. Meanwhile, we experienced other major floods that didn't quite measure up to those disasters in terms of life loss and damage. Stormwater changed too. It mostly seemed under control, but stormwater quality in the U.S. continues as a difficult issue.

In developing countries, flooding and stormwater are persisent and major issues. It is common for heavy rain to paralyze great cities like Jakarta and Manila, which lack adequate drainage systems and combat many land use problems. In Bangladesh, major flooding is a continuing fact of life. Mitigating problems like these will take much more than installation of small drainage systems. Major issues of urbanization and land use control must also be faced.

To illustrate how things come full circle, I recently found myself in the position of the person being regulated in the flood plain. We own a small alfalfa field in a rural area nearby and hope that family members will someday build a home there. The county is urbanizing and commissioned a flood plain study that showed that the road would be flooded and about one-third of our property would be in the 100-year flood plain. The study seemed flawed to me because there is not even a waterway nearby. The problem is what you might call "sheet flooding" due to lack of drainage ways. So, I wrote a long and technical letter about this to the authorities, but my pleas for relief went unanswered. While this is not a big thing, it is an example of "being regulated" and the need for "regulatory reform" (see Chapter 15).

REFERENCES

Akmalah E. (2010). *Integrated Flood Management Model: A Socio-Technical Systems Approach to Overcome Institutional Problems in Jakarta*. Ph.D. Dissertation, Colorado State University, Fort Collins, Colorado.

Debo T. and Reese A. (2003). *Municipal Stormwater Management*, 2nd edn. Lewis Publishers, Boca Raton, FL.

Dubler J. R. and Grigg N. S. (1996). Dam safety policy for spillway design floods. *J. Prof. Iss. Eng. Educ. Pract. ASCE*, October, **122**(4), 163–169.

Federal Emergency Management Agency (2010a). *HAZUS-MH Overview*. http://www.fema.gov/plan/prevent/hazus/hz_overview.shtm (accessed 20 June 2010).

Federal Emergency Management Agency (2010b). *About the National Dam Safety Program*. http://www.fema.gov/plan/prevent/damfailure/ndsp.shtm (accessed 20 June 2010).

Hadimoeljono M. B. (1992). *Flood Control Management in River Basins: Strategies for Integration*. Ph.D. Dissertation, Colorado State University, Fort Collins, Colorado.

Katsuhama Y. (2010). *Capacity Building for Flood Management in Developing Countries Under Climate Change*. Ph.D. Dissertation, Colorado State University, Fort Collins, Colorado.

Platt R. H. (1997). Classics in human geography revisited. *Progr. Hum. Geogr.*, **21**(2), 243–250.

Rahardjo H. A. (2002). *Management of Storm and Flood Control in New Developments in Indonesia*. Ph.D. Dissertation, Colorado State University, Fort Collins, Colorado.

Task Force on the Natural and Beneficial Functions of the Floodplain (June 2002). *The Natural & Beneficial Functions of Floodplains*. US Government Printing Office, Washington, DC.

White G. F. (1945). *Human Adjustment to Floods: A Geographical Approach to the Flood Problem in the United States*. Ph.D. Dissertation, University of Chicago. Also published as Research Paper 29 (1945). Department of Geography.

White G. F. (1969). *Strategies of American Water Management*. University of Michigan Press, Ann Arbor.

Chapter 10

Multipurpose, river basin and regional organizations

A special class of water management organizations has been created to solve some of the complex problems that occur due to the mismatch between political and hydrologic boundaries. These organizations serve across areas such as river basins and other multi-jurisdictional regions and they can integrate services to achieve multiple purpose management goals.

No single name captures the full range of these diverse organizations, but the term "basin organizations" describes much of what they do as in the International Network of Basin Organizations (INBO, 2010), which represents organizations, which in addition to basins, serve multiple purposes over other types of regions. A term such as "multipurpose, basin and other regional organizations" is more descriptive, even if it is awkward.

Regional water organizations have many functions and purposes.

These organizations are difficult to characterize; it's as if each one is cut from a different mold. Still, they perform many essential water services and they provide important but subtle forces for coordination. Governance of these multi-functional organizations is difficult and they often seem to break apart. It is important for water managers to understand the need to support them, even when their programs do not always align with local interests.

INTEGRATED WATER MANAGEMENT ORGANIZATIONS

When water crosses political jurisdictions it creates special difficulties in management, as for example the case of transboundary water issues, which are covered in Chapter 13. In response the water community seeks to integrate its responses to obtain the best overall results. This integrated approach to water management faces barriers, however, from competing political jurisdictions. The organizations in this sector are perhaps unique among public sector management entities in that they exist to provide integrated solutions to these problems.

Looking across other public sector management organizations, such as for transportation, energy, communications, or waste management, you do not find anything quite like these multi-purpose water organizations. Perhaps the closest are the areawide organizations to coordinate planning called Councils of Government or Metropolitan Planning Organizations for transportation.

This chapter describes how multipurpose, river basin and regional organizations are governed, and provides examples from different regions of the world. Although their organizational forms vary, these types of organizations are used throughout the world for many essential water management purposes. It is not easy to create or to sustain their operations, but the success of these organizations underscores how integrated management must be achieved.

THE SECTOR

The sector is difficult to characterize because its organizations take on different forms and have diverse functions. Its organizations are usually either river basin agencies or special purpose agencies at the regional and/or sub-state level. They take on different functions to fulfill their special purposes and serve as planning agencies, river basin authorities, regional multi-purpose utilities and special water management districts.

Whereas water utilities are distinct and relatively easy to characterize with their similar services and defined service areas, river basin and regional service organizations offer varied services and their service areas are defined for each case by legislation, river basins, or broad political jurisdictions, such as counties, multi-county regions, and states. Many of them serve utilities, cities, and industries on a wholesale basis rather than serving residential customers directly, but some of them also operate as utilities or providers of direct water services. Some regional organizations are also involved in flood control.

Most organizations in this sector engage in planning, usually on behalf of multiple entities and across a river basin or other regional area. Many are also involved with project development and/or operation, although some do much more than others. Some have roles in environmental water quality, but few of them have regulatory roles because a main purpose of regionalization in water management is to coordinate and serve, rather than to regulate.

While the organisations are difficult to classify, I placed them into three general groups:

- Regional organizations focused on planning and areawide services (usually basin organizations with combinations of planning, regulatory, and service activities)
- Regional organizations focused on local services (usually regional utilities and special districts focused on urban services)
- Federal organizations with service missions

In addition, international river basin organizations provide venues for national authorities to meet and negotiate on water management. These do not provide services or plan projects, except in rare circumstances. They will be discussed briefly toward the end of the chapter.

REGIONAL ORGANIZATIONS FOCUSED ON PLANNING AND AREAWIDE SERVICES

These organizations are usually basin organizations with combinations of planning, regulatory, and service activities. They may have other geographical boundaries, such as combinations of local or regional, sub-state governments.

Classifying the organizations

These organizations are difficult to classify rigorously. Priscoli (2009) and Kenney and Lord (1994) studied groups of river basin organizations and focused on how they were governed as:

- Single federal administrator (examples, Colorado River Compact and Columbia River Treaty Organization)
- Inter-Agency-Interstate Commissions (examples, river basin commissions under the Water Resources Planning Act of 1965, called Title II commissions)
- Federal Interstate Commission (example, Delaware River Basin Commission)
- Interstate Compact Commission (Potomac River)
- RBOs as Boundary Organizations (examples, International Joint Commission and the International Boundary Waters Commission)

They included the Tennessee Valley Authority, but it is in a class of its own. Hooper (2005) cited 26 examples of river basin organizations that ranged from small watershed associations to large international commissions, such as the International Commission for the Protection of the Danube River. Some examples focused on areas other than river basins.

The U.S. has areawide organizations to coordinate intergovernmental planning, but regionalism in water management has proved to be a challenge. Examples of areawide planning organizations are the Councils of Government (COGs) for multiple services and the Metropolitan Planning Organizations (MPOs) for transportation. The COGs are coordinated through the National Association of Regional Councils (NARC) and the through the Association of Metropolitan Planning Organizations (AMPO). While the COGs might consider water services in their planning, their authorities in water are limited.

To recognize the need for regionalism in water planning, the U.S. initiated a national program through the Water Resources Planning Act of 1965 (WRPA). However, the program failed and now there is no coordinated national effort for regional water planning. However, a number of planning bodies that have been created by ad hoc legislation are in operation. If there is any governance lesson, it is that it is difficult to enact a top-down program such as the river basin commissions required by the WRPA.

River basin organizations seem well-established in Europe, perhaps better than in the U.S. due to the small sizes of some countries and the greater need for international cooperation. Currently, the EU Water Framework directive requires river basin planning and organization. The UK's system of regional water organizations and now companies aligns with river basins. France has a river basin system with arbiter role between polluters and users and with investment incentives. Germany has a long tradition in the Ruhr Basin. Portugal and Spain have long tradition of river basin organizations mainly focused on supply. In Russia, 17 river basin agencies operate under the Ministry of Natural Resources, with five of them on the Volga River (Priscoli, 2009).

The next section presents examples of regional organizations that are focused on planning and areawide services.

Regional organizations focused on planning and areawide services

Mississippi River Basin Commission

One of the planning bodies formed by ad hoc legislation is the federally-sponsored Mississippi River Commission (MRC), which dates back to 1879 with a focus on navigation and flood control of the river. This commission is dominated by government control, as is obvious from the military representation. Its members are nominated by the President and confirmed by the Senate. The commission members are three U.S. Army Corps of Engineers officers, one member of the National Oceanic and Atmospheric Administration, and three civilians, two of whom must be civil engineers. The Commission president is the commanding officer of the

Corps of Engineers Mississippi Valley Division and the other two Corps members are the commanding officers of the Northwest Division and the Great Lakes and Ohio River Division.

Most other U.S. interstate water commissions have an interstate compact to guide them, but the MRC was created to coordinate planning and policy on the Mississippi, Missouri, and Ohio Rivers. Thus, navigation and flood control are important purposes for it.

Upper Colorado River Commission

The Upper Colorado River Commission (UCRC) coordinates the upper basin states (Colorado, New Mexico, Utah, and Wyoming) to use their shares of the Colorado River while meeting constraints posed by legislation. This interstate administrative agency was created when Congress ratified the 1948 Upper Colorado River Basin Compact. The UCRC makes findings of upper basin water use, deliveries at the Lee Ferry gage, and the possible necessity to curtail use to comply with the 1922 Compact. This commission therefore has a limited mission that focuses on interstate water allocation (Colorado River Water User's Association, 2010).

Delaware River Basin Commission

Although it is authorized by an interstate compact, the multi-state Delaware River Basin Commission (2010) has a broader mission than most compact commissions. The mission covers river basin planning, water quality protection, water supply allocation, regulatory permitting, water conservation, watershed planning, drought management, flood loss reduction, and recreation. It involves four states (New York, New Jersey, Delaware, Pennsylvania) and the federal government.

The Commission was created in 1961 when four governors and President John F. Kennedy signed legislation to create it as a regional compact commission with a unified approach to managing a river system. Members of the commission are the four state governors and the federal representative, which is the Division Engineer, North Atlantic Division, U.S. Army Corps of Engineers. The commission is funded by the states and federal government, fees, water use charges, and fines, and grants.

South Florida Water Management District

The water management districts in Florida are a good example of how a state government can create regional authorities for special purposes. Florida has five water management districts, organized with both river basin and areawide boundaries. They administer flood protection programs, perform technical investigations, develop plans for drought response, manage lands, and administer delegated regulatory programs that include consumptive use management, aquifer recharge, well construction and surface water management, including stormwater programs (Florida Department of Environmental Protection, 2010).

The South Florida Water Management District (SFWMD) is the largest of the five water districts and was created in 1949 to focus initially on flood control. Its original name was the Central and Southern Florida Flood Control District, which focused on managing a flood control project. The other water management districts were created in 1972, with the Florida Water Resources Act. The SFWMD operates in 16 counties in the southern half of the state and serves a population of some 7.5 million. Currently, it has a key initiative for restoration of the Everglades, which is the largest environmental restoration project in U.S. history.

Trinity River Authority

The Trinity River Authority (2010) in Texas serves a 715-mile long river corridor that passes through two major population centers (Dallas/Fort Worth and Houston) and ends at the Galveston Bay and estuary system. The Authority is a political subdivision of the state that was created by the Legislature in 1955 for three functions: basin master planning, serving as local sponsor for federal water projects, and providing areawide services for wastewater treatment, water treatment, recreation and reservoir facilities. It also manages tax-exempt financing to assist municipal water and wastewater and industrial air and water pollution control projects. The Authority operates as a government-owned utility and is entirely self-funded from revenues from services provided by its projects. It is governed by a Board of Directors that is appointed by the Governor with the advice and consent of the Senate.

Ohio River Sanitation Commission

The Ohio River Valley Water Sanitation Commission (ORSANCO, 2010) was established by compact in 1948 to represent eight states and the federal government in pollution control activities in the Ohio River Basin, including tributaries. Its programs include wastewater standards, biological assessments, monitoring, and special surveys and studies. ORSANCO also coordinates emergency response and promotes public participation.

ORSANCO's membership comprises three commissioners from each State and three representing the federal government. Each state can decide how to appoint its commissioners and the federal representative is appointed by the President or in another manner as Congress might choose. The commission submits its budgets and annual reports to the state governments.

Ruhr River Association

The Ruhrverband (Ruhr River Association, 2010) would seem to have similarities to U.S. water quality authorities, but it actually has wider powers and has existed longer. It is one of some eleven water associations in North Rhine-Westphalia, a state of Germany. These associations cooperate to achieve regional and inter-regional water management goals for a densely populated area. The Ruhrverband focuses on water quality with a network of wastewater treatment plants that has operated for more than 90 years. It also owns and operates a series of reservoirs.

Governance of the Ruhrverband, which is incorporated under German public law, extends back to authorities under Prussian law. It is a self-governing body subject to oversight by the federal state of North Rhine-Westphalia and regulated by the Ministry of Environment, Nature, Agriculture and Consumer Protection. Members include the communities and districts in the basin and the wastewater dischargers, as well as water users and owners of hydropower plants.

Since 1990 the Ruhrverband has been organized like a public limited company. Decision-making is by an assembly of associates, a supervisory board and an executive board. The some 152 delegates on the assembly of associates include 150 from member organizations or elected members and two from the Chambers of Agriculture. The concept of self-administration of the association is based on the polluter–pays principle within the basin.

Colombia: Regional Autonomous Corporations

In Colombia, the Regional Autonomous Corporations (CARs) have pivotal roles in water management. Whereas other Latin American countries mostly take top-down national approaches to water management, Colombia took a regional approach beginning in the 1950s when it created the Regional Autonomous Corporation of the Cauca Valley (CVC, 2010). CVC was created in 1954 for economic development and modeled somewhat after the Tennessee Valley Authority. The creation of CVC is regarded as the first experiment with decentralized government in Colombia. It has accomplished large water projects such as reservoir and hydroelectric construction, as well as other environmental achievements.

Between 1960 and 1988, 18 additional CARs were created, but they ceased to follow watershed boundaries and took on diverse regional duties that went beyond water management. The 1991 Constitution created the Regional Autonomous Corporation of the Magdalena River with the specific purpose of water resource development. They were recast as environmental management authorities and additional CARs were established to work with the national environmental agency on watershed planning and other resource issues.

Resources for the Future Inc., a Washington-based research institute, performed an institutional analysis of the CARs (Blackman *et al.*, 2006). Their findings resonate as similar to those you might find anywhere that regional governments overlap with other jurisdictions. They thought that the CARs did not exhibit uniform performance, which seemed to be correlated with institutional experience, geographic size, and level of poverty. The human, technical and financial capacities of the CARs to manage varied markedly and political patronage was a factor in staffing. Another political issue was excessive influence of certain interest groups. As is often the case in regional government, financial and jurisdictional conflicts arose between the CARs and other government units with environmental responsibilities, such as Urban Environmental Authorities and municipalities. Coordination between the CARs and the national environmental agency also seemed inadequate. Environmental enforcement seemed to be a problem, and voluntary regulation used by the CARs was often rated as ineffective. The joint roles of CARs in managing and regulating sanitation infrastructure created conflicts of interests.

Summary of the organizational types

Table 10.1 gives a summary of these types of organizations and shows their variation:

Table 10.1 Organizational types and variations

Organization	Services	Boards of Governors
Mississippi River Basin Commission	Navigation and flood control	Federal appointment
Upper Colorado River Commission	Interstate water entitlements	State representatives
Delaware River Basin Commission	Multipurpose, focus on basin planning, water quality, water allocation, flood loss reduction	Four state governors and federal representative (from USACE)
South Florida Water Management District	Multipurpose, focus on flood protection, drought response, delegated regulatory programs	Elected board
Trinity River Authority	Basin planning, sponsor for federal projects, areawide services	Board of Directors appointed by the Governor
Ohio River Valley Water Sanitation Commission	Wastewater standards, assessments, monitoring and special studies	Commissioners from each state and federal government
Colombia Regional Autonomous Corporations	Regional environmental management and economic development	Government appointments

Regional organizations focused on services

Globally, many regional organizations exist to provide water-related services within basins.

Tennessee Valley Authority

Perhaps the most famous U.S. model of a regional organization with a service mission is the Tennessee Valley Authority (TVA, 2010a), which was created during the New Deal when President Roosevelt asked Congress to create "a corporation clothed with the power of government but possessed of the flexibility and initiative of a private enterprise." Congress passed the TVA Act in 1933 and the Authority has operated since then as a government–sponsored enterprise that provides electric power and water management in the Tennessee River valley.

The TVA was started as one of President Franklin Roosevelt's New Deal programs.

Figure 10.1 shows the Tennessee Valley region that has been transformed through the activities of TVA over some 70 years (TVA, 2010b). It is governed by a nine-member Board of Directors who are nominated by the President and confirmed by the Senate for five-year terms. Although TVA is a government corporation, the Consolidated Appropriations Act of 2005 changed its management structure and required it to file financial reports with the Securities and Exchange Commission.

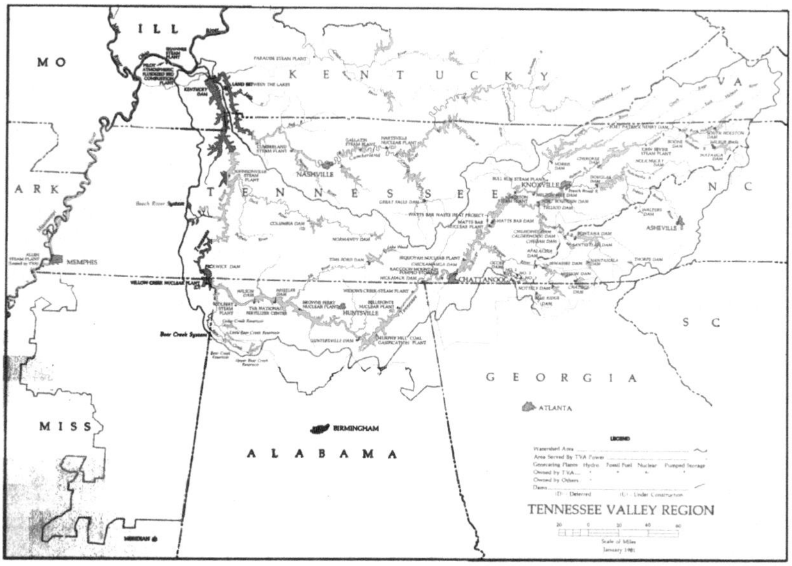

Figure 10.1 Tennessee River Region [Tennessee Valley Authority (1983). A History of the Tennessee Valley Authority: 50th Anniversary Edition, Knoxville, Tennessee]

Salt River Project

Another regional utility with government sponsorship is the Salt River Project (SRP, 2010), which was the nation's first multipurpose reclamation project in 1903 as authorized under the National Reclamation Act. Unlike TVA, it has largely cut its government control links and is now the third-largest public power utility in the U.S. and one of Arizona's largest water suppliers with a 2,900-square-mile service territory in central Arizona.

The SRP evolved from the Salt River Valley Water Users' Association, which was created by ranchers and farmers in metropolitan Phoenix to manage water rights and settle disputes. Today, the Association is a private corporation and part of the SRP, which also includes the Salt River Project Agricultural Improvement and Power District, a political subdivision of Arizona. SRP operates some of its facilities under contract with the U.S. Bureau of Reclamation and other facilities are owned outright. The president of SRP is the chief executive office and chairman of the Board of each organization. The SRP president, vice president, Boards and Councils are elected by a proportional system based on land ownership. This concept stems from an original financial commitment of landowners who pledged land as collateral for the federal loan to build the original Roosevelt Dam in the early

1900s. Four at-large Board members are elected on a one-landowner, one-vote basis. The constitutionality of this acreage-based voting system was tested and upheld by a U.S. Supreme Court case in 1981.

Metropolitan Water District of Southern California

The Metropolitan Water District of Southern California (2010), or the "Met," is the largest regional wholesale water provider in the U.S. It is a consortium of 26 cities and water districts that provides drinking water to nearly 20 million people in Southern California. It was formed in 1928 under a special bill of the California Legislature with approval of voters in the participating jurisdictions. Its first main facility was the Colorado River Aqueduct, which was completed in 1941. Later, it also received water from the California State Water Project. It has a 5,200 mile service area and delivers an average of 1.7 BGD (2,630 cfs). The Met is governed by a board of directors with members from the participating water providers.

Tampa Bay Water

Tampa Bay Water (TBW, 2010) is another regional wholesale water provider, located in Florida. It was created in 1998 through contracts and legislation that converted the West Coast Regional Water Supply Authority to a new form that ended some conflicts and created an alliance between six city and county governments in west-central Florida. The problem with the West Coast Regional Water Supply Authority was that it was a cooperative and had inherent financial inequities. This created a situation where some members would try to buy water only in their own interests and the cost of water from the Authority would vary among members. A majority vote of the Board was not always binding and a large member could withhold funding for joint projects. The concept of a regional was explored by a policy panel comprised of the Regional Water Supply Authority Board and administrators and utility directors for the member governments. The Florida Legislature directed the Authority to evaluate its operation and make recommendations for improvements, thus leading to the establishment of TBW in its present form. Adams (1998) explained details of the conflicts and how they were resolved in a study at Colorado State University.

Lake Constance Water Supply Association

In Germany, the Bodensee-Wasserversorgung Public Corporation (2009), also called the Lake Constance Water Supply Association, is the largest long-distance water supplier in Germany and meets needs of some 180 communities and member associations. Figure 10.2 shows the route of its water supply system, beginning at Lake Constance (Bodensee) and continuing northward to supply its member utilities, including Stuttgart.

It is similar to the Met in that it is controlled by the members and provides regional water supply services. The history of the corporation dates back to 1871, when discussions about long distance water transport began and by 1912 efforts were underway to have an association for water transport. After World War II, several water emergencies in Stuttgart and other communities occurred, especially during the drought years 1947 and 1949. A study commission reported out in 1953 and communities were given the opportunity to join a public corporation. In a few years, the facilities began to come on line, leading to today's advanced system (Melhorn and Geckeler, 2005).

East Bay Municipal Utility District

The East Bay Municipal Utility District (EBMUD, 2010), a large, multi-jurisdictional municipal utility district is a water and wastewater management utility serving a population of about 1.3 million in the the Bay Area of California. The State of California authorized municipal utility districts through legislation in 1921 and, EBMUD was organized in 1923. In 1929 its Pardee Dam was completed, and was the highest dam in the world at the time. In 1951 its wastewater treatment system came on line. In 1994 the district implemented a Seismic Improvement Program. It has a staff of nearly 2000 and a seven-member Board of Directors who are elected from wards within the service area.

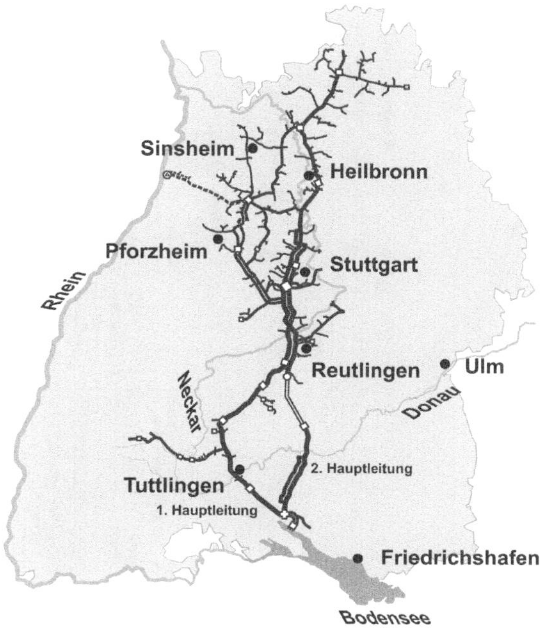

Figure 10.2 Lake Constance Water Supply System [Zweckverband Bodensee-Wasserversorgung (Lake Constance Water Supply Association, Stuttgart, Germany]

Imperial Irrigation District

The Imperial Irrigation District (IID, 2009) is the largest combined water and electric power district in the U.S. IID was established in 1911 under the California Irrigation District Act. IID is the largest irrigation district in the U.S. and its electrical division, IID Energy, is the sixth largest consumer-owned utility in California. The district delivers water to over 450,000 acres of farmland in Southern California. It is governed by an elected five-member Board.

Central Nebraska Public Power and Irrigation District

The Central Nebraska Public Power and Irrigation District (CNPP&ID, 2010) is another combined water and electric power district. It was created as a political subdivision of Nebraska during the Great Depression but it might not have survived without financial assistance from the New Deal's Public Works Administration that funded a "Tri-County Project." Its s hydroelectric facilities are licensed by the Federal Energy Regulatory Commission and it has recently gone through a difficult re-licensing process with many environmental issues to resolve. The District is governed by an elected 15-member board of directors serving six-year terms and drawn from the regions that it serves.

Northern Colorado Water Conservancy District

The Northern Colorado Water Conservancy District (NCWCD or "Northern", 2010) is mainly a wholesale water authority that was established under Colorado's Water Conservancy District Act of 1937. As Figure 10.3 shows, NCWCD transfers water across the Continental Divide of the U.S. from the Colorado River Basin to a network of irrigation and municipal users on the East Slope of the Rocky Mountains.

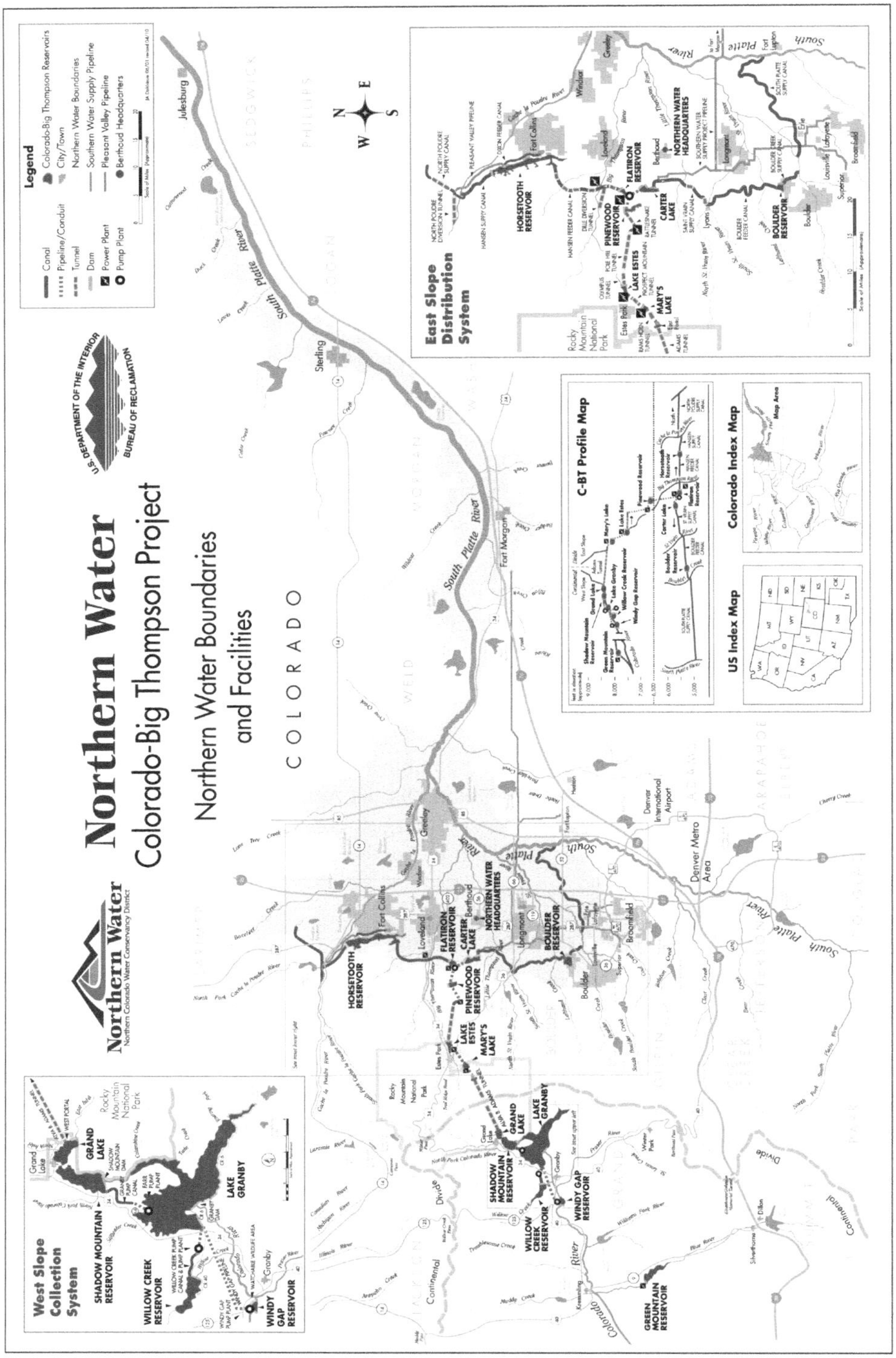

Figure 10.3 Northern Colorado Water Conservancy District (Northern Colorado Water Conservancy District, Berthoud, Colorado)

Northern operates facilities owned by the Bureau of Reclamation and owns water rights to distribute to farm and municipal water users. It is governed by a 12-member Board whose members are appointed by district court judges in the individual counties. It has a Municipal Subdistrict, which elects its slate of officers but has the same Board as the District. Its method of selecting directors is often questioned by environmental interests who advocate direct election of directors.

Urban Drainage and Flood Control District

After disastrous 1965 flooding, the State of Colorado decided to organize a regional flood control district. The organizational meeting of the Urban Drainage and Flood Control District (UDFCD) was on July 28, 1969 and the district boundaries consisted of five counties containing some 30 cities and towns and the City and County of Denver. Initially, the district concentrated on planning, and by now has planned many miles of major drainageways. Early results showed the need for a floodplain management program and many miles of floodplains have been delineated so local governments can use the information for land use decisions. The district's is funded by a property tax to provide funds for design and construction and operates a 50–50 matching program with local governments. Maintenance proved to be a problem and in 1979 the district was authorized by the legislature to levy an additional tax for maintenance. The district now has four active programs: master planning, floodplain management, capital improvement, and maintenance. Then it added a South Platte River maintenance program. The district entered the water quality field when stormwater quality became an issue, beginning in the 1970's. It is governed by a Board drawn from participating local governments in the Denver region (Grigg, 1996).

Summary of organizational types

While these organizations share the attribute of being regional service providers, they differ significantly in ownership, types of services, and governance. Table 10.2 summarizes them.

Table 10.2 Organizational types and differences

Organization	Services	Governance
Federally-owned organization: Tennessee Valley Authority	Electric power and water management	Nine-member Board of Directors, nominated by President and confirmed by the Senate
Mixed public and private organization: Salt River Project	Public power and water supplier	Officers and Board elected by proportional system based on land ownership
Regional urban water wholesalers: 1. Metropolitan Water District of Southern California 2. Tampa Bay Water	Wholesale water providers	Board members are designated by participating local governments
Regional water and wastewater utility: East Bay Municipal Utility District	Water and wastewater management	Board of directors elected.
Electric power and irrigation districts: 1. Imperial Irrigation District 2. Central Nebraska Public Power and Irrigation District	Water and electric power	Elected board.
Regional irrigation water provider: Northern Colorado Water Conservancy District	Wholesale water for irrigation and conversion to urban uses	Board appointed by district court judges
Regional drainage and flood control district: Urban Drainage and Flood Control District	Planning, management of major drainageways	Board drawn from local governments

NATIONAL PLANNING AND SERVICE AUTHORITIES

Federal organizations with service missions

In the previous section, TVA was explained as a federal government-owned river basin organization, but it is really a service provider, not a river basin planning organization. In any case, it was created in the 1930s and could not be created in today's political climate.

However, two other government–owned water providers are active, the U.S. Bureau of Reclamation (USBR) and U.S. Army Corps of Engineers (USACE). These national agencies own and operate dams and act as regional water providers.

In many countries, it is common for national organizations to have responsibilities for water management. Asia has a network of Network of Asian River Basin Organizations (2010), which provides some coordination among organizations with similar missions. It was established at the 2003 World Water Forum and has a secretariat at the Japan Water Agency. For example, in Korea, K-Water (former Korea Water Corporation), has extensive water management responsibilities including regional water management. China has seven major river basin commissions, but they are extensions of the Ministry of Water Resources, rather than creatures of lower level government agencies. The Yellow River Conservancy Commission is the most active (Turner, 2007). Indonesia has river basin organizations for the Jatiluhur Water Authority and the Brantas Water Management Corporation.

Another regional authority with a mission similar to TVA for economic development is the Mahaweli Authority of Sri Lanka (2010), which is a multi-purpose organization with a focus on integrated rural development and based on the resources of the Mahaweli and six other river basins. The objectives are to increase agricultural production, hydropower, employment, flood control, and settlement of landless poor. Investigations leading to establishment of the system and the Authority began with investigations in the 1950s and the Authority was incorporated in 1979.

Nearby is Australia's Murray-Darling Basin Authority (2010), which replaced the Murray Darling Basin Commission. The river basin includes The catchment areas for the Murray and Darling rivers and their tributaries, a total of 23 river valleys and an area over 1 million square kilometers or 14% of Australia. In addition to serving metropolitan areas of New South Wales and Victoria, the basin is very important to agriculture, with nearly 40% of the national agricultural production located in it. The basin has experienced severe drought problems and the Authority has developed a number of response mechanisms.

In Brazil, the Rio Sao Francisco Development Agency and SUDENE (organization for planning and development in the Northeast) are river basin and areawide water management organizations. Mexico currently has six basin commissions. In Jordan, the Jordan Valley Authority exercises multiple purpose water management responsibilities.

In Africa, Nigeria has 11 basin authorities, including the Niger Delta Development authority. Kenya has three basin authorities.

INTERNATIONAL MULTI-LATERAL RIVER ORGANIZATIONS

A number of international river multi-lateral treaty organizations provide venues for country representatives to negotiate on water management. Examples described in the next few paragraphs.

The Danube River basin touches 19 countries, and 13 of them are parties to the Danube River Protection Convention, which was established in 1998. The Basin was the first Regional Programme funded by the

Global Environment Facility (GEF), which was itself established in 1991. The United Nations Development Programme (UNDP) has implemented GEF projects that support the Danube. The International Commission for the Protection of the Danube River (ICPDR, 2010) was also formed in 1998 and is carrying out programs funded by GEF/UNDP. These arrangements show the need for involvement of international institutions to facilitate negotiations toward improved water management in complex international settings.

The international institution for the management of the Rhine River is the International Commission for the Protection of the Rhine (ICPR), which was founded 1950 by Switzerland, France, Luxembourg, Germany and the Netherlands. The idea was to create a forum to discuss and solve problems of water management, especially flooding and pollution. The representatives of member states govern the commission, but decisions taken by it do not bind the member states. The commission has a small secretariat and works mostly through a Coordination Group (The Water Page, 2010).

The Mekong River Commission (MRC, 2010) was formed in 1995 by Cambodia, Lao PDR, Thailand and Viet Nam, when the four countries signed an agreement and agreed on joint management of the river. Its organization was supported by studies dating from 1957 under the UN's Mekong Committee. China and Myanmar are Dialogue Partners of the MRC. The Mekong River Commission Secretariat is based in Vientiane, Lao PDR, and supports the MRC Council and Joint Committee.

The Nile Basin Initiative (2010) is the current vehicle to work toward permanent cooperation on the Nile River Basin, which involves Burundi, D.R. Congo, Egypt, Ethiopia, Kenya, Rwanda, Sudan, Tanzania, and Uganda.

In Latin American, the Coordinating Intergovernmental Committee (CIC) was established for the La Plata Basin, which helped prepare the treaty of La Plata Basin. The La Plata River basin involves Argentina, Bolivia, Brazil, Paraguay, and Uruguay and drains the Parana, Paraguay and Uruguay river systems. It includes Pantanal, which is the largest wetland in the world. The 1969 La Plata River Basin Treaty is the basic multilateral agreement, which provides the framework for negotiations. Other agreements have involved large and controversial projects, such as the construction of Itaipu Dam and the current proposed "Hydrovia" or river navigation project, which would affect the Pantanal (Wolf and Newton, 2010).

The International Joint Commission (2010) was established by the 1909 Boundary Waters Treaty, and has three U.S. members appointed by the President and three Canadian members appointed by the Governor in Council on the advice of the Prime Minister. It has joint expert boards to help it carry out its responsibilities.

The Great Lakes Commission (2010) is an interstate compact agency for water management in the lakes and St. Lawrence River. Its members are eight Great Lakes states and Ontario and Québec as Associate Members. Governance is by representatives appointed by the participating jurisdiction. The Commission was established by the states in 1955 and recognized by Congress in 1968. The international component was established in 1999 through a Declaration of Partnership.

GOVERNANCE OF MULTIPURPOSE REGIONAL ORGANIZATIONS

In the cases of direct water services, governance is concerned with policy, empowerment, and control. However, in the case of multipurpose, basin and other regional organizations, governance must deal both with the purposes of the organizations and with their special class of operations, which can be more complex than those of single-purpose organizations.

To explain these two channels of governance, I will draw from the previous section and its examples of organizations.

Policy

In the U.S. governmental system, the term "special purpose" organization was adopted to explain independent, special-purpose governmental units that exist as separate entities with substantial administrative and fiscal independence from general–purpose local governments (U.S. Bureau of Census, 2002). They are needed to provide specific services not available from existing general–purpose governments. They are usually called districts or authorities and are distinct from organizations with similar names that are subordinate units to general-purpose governments.

This sector usually draws its organizational and business policy from state laws, but multi-state and/or federal organizations draw their policy from national legislation. Any of these organizations also draw policy from the sectors that correspond to its functions. For example, a regional wholesale water supplier would be authorized by state legislation and follow national policy for source water protection.

Empowerment

Institutional empowerment for these special purpose organizations derives from their enabling legislation and their mandates specify how they are to be financed. As the organizations can be fragile politically, their empowerment might require a firm commitment from the sponsors to keep them from falling apart due to polarizing political conflicts.

Control

Given their special purpose nature, control of the organizations in the sector comes via the services they perform, and in some cases they have regulatory powers themselves.

As many of these organizations serve general purpose governments, they are likely to have corporate governance structures that include representatives of the participating units.

CONCLUSIONS

Multipurpose, river basin and regional organizations offer intriguing possibilities to integrate water management across differing purposes, regions and stakeholders. They provide institutional mechanisms to cut through the barriers presented by balkanized government jurisdictions. It is difficult to characterize them because they are organized to respond to unique local and regional situations. However, they share the attribute of offering planning and coordination across diverse interests.

Although they offer attractive possibilities, the governance issues of multipurpose, river basin and regional organizations can be an order of magnitude more difficult than those of single purpose organizations with limited service areas. It is easier for the single purpose organizations to focus on their customers and missions that it is for organizations with many more stakeholders. That is not necessarily bad news because the multi-stakeholder organizations have usually been created to deal with coordination or sharing issues.

The degree of difficulty in managing and governing these organizations depends on the politics involved. The difficulty can sometimes be mitigated when there is a strong shared interest in improvement, such as in water quality in the Great Lakes. When stakeholders perceive win-lose scenarios, the difficulties can be much greater, as in the case of allocating water across boundaries.

International and interstate politics can trump achievements in cooperation and coordination that have been negotiated by water managers. Shared governance is an attractive concept, but it can seem idealistic when diverging political interests drive wedges between negotiators.

Much more study is needed of multipurpose, river basin and regional organizations to demonstrate their benefits and how governance can be arranged to facilitate their application in areas of need.

Postscript

Although it's true that IWRM is a tough challenge, the multipurpose, regional organizations introduced in this chapter are bright lights that offer ways to integrate solutions and reconcile differences. I've been intrigued by them since the 1970s, when I had my first opportunity to serve as a consultant with a California water district, which focused on regional water management and flood control. At the time I did not realize there were so many different types of regional organizations, but I had already noted their possibilities in transportation, regional services and stormwater. Now, I like to promote them as a good way to overcome local constraints in working toward Total Water Management, but it's difficult because of many political issues.

I've already noted the many advantages wrought by Denver's Urban Drainage and Flood Control District, and I've paid special attention to its growth and development due to a friendship with their long-serving Executive Director, Scott Tucker. Scott assumed that position in 1972 and led the organization as it developed for over thirty years. He was twice-honored as a "Top Ten" public works director by the American Public Works Association for his leadership.

The work of river basin commissions came onto my radar screen in the 1970s when, among others, I had responsibility for North Carolina's water resources planning program. This meant I was a member of a couple of commissions myself and supervised the staff that carried out basin studies under the Water Resources Planning Act. I noticed that when planning was carried out top-down, it mainly failed. This was particularly the case for a federally-sponsored river basin study that we were conducting at the time. When planning studies responded to deeply-felt local needs, they had a better chance to succeed.

During the 1980s and 1990s I began to visit Asia and I noticed that both Japan and China had long histories of regional water organizations, albeit of different types. Japan seemed to work on regional problems through its centralized political system and to like public-private and national-local cooperative organizations. This seemed to grow out of their cultural affinity with cooperation. China had vast problems like the scale of their country and had a number of functional river basin commissions, such as the Yellow River Conservancy Commission.

During the 1990s I joined the Board of Directors of a local organization named the Big Thompson Water Forum. It was a private volunteer organization that monitored and studied the watershed with an eye toward improving water quality. While in Colorado it is well-known that water rights cause conflict, I think we were unaware of some of the water quality conflicts that can arise. It's a long story, but the Big Thompson River carries a lot of interbasin transfer water from one side of the Rocky Mountains to the other. Due to land use changes in the basin of origin, the water quality in the receiving basin is affected and this requires a difficult planning process with a lot of players.

I described Colombia's autonomous regional corporations earlier, and when studying them I was struck by the foresight of the policy planners in establishing them. They seem to offer a good way to accomplish regional goals while implementing a responsible approach to environmental management. However, they also face many constraints.

At the end of the day, it seems to me that managing water through these organizations is like a lot of other regional approaches to problem-solving: easy to say, hard to do.

REFERENCES

Adams A. (1998). *Analysis of Regional Water Conflicts: The Case Study Approach*. Ph.D. Dissertation, Colorado State University, Fort Collins, CO.

Blackman A., Morgenstern R. and Topping, E. (2006). *Institutional Analysis of Colombia's Regional Autonomous Corporations*. Resources for the Future, Washington, DC.

Bodensee-Wasserversorgung Public Corporation (2009). *Water Moves Us*. http://www.zvbwv.de/ (accessed 19 December 2009).

Central Nebraska Public Power and Irrigation District (2010). *Operations of the Central District*. http://www.cnppid.com (accessed 4 April 2010).

Colorado River Water User's Association (2010). *About Us*. http://www.crwua.org (accessed 3 April 2010).

Corporación Autónoma Regional del Valle del Cauca (2010). *Asi es la CVC*. http://www.cvc.gov.co/vsm38cvc/ (accessed 4 July 2010).

Delaware River Basin Commission (2010). *Who We Are*. http://www.state.nj.us/drbc/ (accessed 3 April 2010).

East Bay Municipal Utility District (2010). *Our History*. http://www.ebmud.com/ (accessed 4 April 2010).

Florida Department of Environmental Protection (2010). *Water Management Districts*. http://www.dep.state.fl.us/secretary/watman/ (accessed 12 January 2010).

Great Lakes Commission (2010). *About the Great Lakes Commission*. http://www.glc.org/ (accessed 16 May 2010).

Grigg N. (1996). *Water Resources Management: Principles, Cases and Regulations*. McGraw-Hill, New York.

Hooper B. (2005). *Integrated River Basin Governance: Learning from International Experience*. IWA Publishing, London.

Imperial Irrigation District (2009). *About IID*. http://www.iid.com/About/AboutIID (accessed 9 December 2009).

International Commission for the Protection of the Danube River. *15 Years of Managing the Danube Basin*. http://www.icpdr.org (accessed 4 April 2010).

International Joint Commission (2010). *Who We Are*. http://www.ijc.org (accessed 16 May 2010).

International Network of Basin Organizations (2010). *The Website on Basin Management in the World*. http://www.inbo-news.org/ (accessed 10 May 2010).

Kenny D. S. and Lord W. B. (July 1994). *Coordination Mechanisms for the Control of Interstate Water Resources: A Synthesis and Review of the Literature*. Task 2 Report for the ACT-ACF Coordination Mechanism Study. Institute for Water Resources, USACE, Alexandria, VA.

Mahaweli Authority of Sri Lanka (2010). *About Us*. http://www.mahaweli.gov.lk (accessed 4 July 2010).

Mekong River Commission for Sustainable Development (2010). *The Mekong River Commission*. http://www.mrcmekong.org/ (accessed 15 May 2010).

Melhorn H. and Geckeler A. (2005). 50 Jahre Bodensee-Wasserversorgung. Ein hervorragendes Beispeil fuer kommunale Azusammenarbeit und Solidaritaet. *Wasser-Abwasser*, **146(**1), 26–32.

Metropolitan Water District of Southern California (2009). *About MWD*. http://www.mwdh2o.com/ (accessed 9 December 2009).

Mississippi River Commission (2010). *About the MRC*. http://www.mvd.usace.army.mil/mrc/about/index.php (accessed 11 January 2010).

Murray-Darling Basin Authority (2010). *About the Basin*. http://www.mdba.gov.au/water/about_basin (accessed 4 July 2010).

Network of Asian River Basin Organizations (2010). *What's NARBO?* http://www.narbo.jp/ (accessed 3 July 2010).

Nile Basin Initiative (2010). *About Us*. http://www.nilebasin.org (accessed 15 May 2010).

Northern Colorado Water Conservancy District (2010). *About the District*. http://www.ncwcd.org/ (accessed 4 April 2010).

Ohio River Sanitation Commission (2010). *About us*. http://www.orsanco.org/ (accessed 4 April 2010).

Priscoli J. D. (2009). *River Basin Organizations*. Appendix to database listing found at Oregon State University. Program in Water Conflict Management and Transformation (PWCMT). http://www.transboundarywaters.orst.edu/ (accessed 3 December 2009).

Rhurverband (2010). *Organisation*. http://www.ruhrverband.de (accessed 15 May 2010).

Salt River Project (2010). *SRP Governance*. https://www.srpnet.com (accessed 4 April 2010).

South Florida Water Management District (2009). *About Us*. http://www.sfwmd.gov/portal/page/portal/sfwmdmain/about%20us (accessed 9 December 2009).

Tampa Bay Water (2010). *Governance Background*. http://www.tampabaywater.org (accessed 4 April 2010).

Tennessee Valley Authority (2010). *Corporate Governance*. http://www.tva.gov/ (accessed 4 April 2010).

The Water Page (2010). *Rhine River*. http://www.africanwater.org/rhine_main.htm (accessed 15 May 2010).

Trinity River Authority (2010). *About Us*. http://www.trinityra.org/ (accessed 12 January 2010).

Turner J. L. (2007). *River Basin Governance in China*. Wilson Center, China Environment Series 7, Washington DC.

U.S. Bureau of Census (2002). *2002 Census of Governments: Volume 1, Number 1, Government Organization*. http://www.census.gov/prod/2003pubs/gc021x1.pdf (accessed 10 December 2009).

Wolf A. T. and Newton J. T. (2010). *Case Study Transboundary Dispute Resolution: The La Plata Basin*. www.transboundarywaters.orst.edu (accessed 15 May 2010).

Chapter 11

Water governance for the triple bottom line

A main shortcoming of unbridled capitalism is its neglect of the social and environmental needs of society. On the other hand, socialism works against the entrepreneurial spirit that raises our standard of living. It seems there is no perfect solution to meeting society's needs. What are we to do?

One bright light in this dilemma is government's tendency to use "triple bottom line" (TBL) analysis to weigh the impacts of its actions in the economic, social and environmental spheres. This enables government to use its resources optimally to further society's goals while safeguarding us against a range of threats. This is not easy and there are many questions about how to identify and evaluate goals and impacts across the categories. Still, this admission by government and citizens that social and environmental issues rank with economic ones lends optimism to our concern about a sustainable future. This chapter is about how the TBL concept can be used in water governance.

A sustainable future will have a positive Triple Bottom Line.

THE TRIPLE BOTTOM LINE CONCEPT

There is much interest today in finding sustainable solutions to urbanization and economic development, and many public organizations are seeking TBL methods to account for their actions. The TBL approach takes note of economic, environmental and social impacts of actions, as opposed to only financial and operating results. Some TBL impacts involve values that are not reflected by the market and are difficult to measure or to finance through the levy of service-based fees. Regardless, social and environmental needs such as safe drinking water for vulnerable populations and sustaining natural ecosystems require special attention from water governance.

The diverse social and environmental purposes of water use place the spotlight on the attribute of water as a public and private good. If water is purely a private good, then neither the environment nor low-income people

 ISBN: 9781843393467. Published by IWA Publishing, London, UK.

can afford enough high-quality supplies. If it is purely a public good, then society has an obligation to provide adequate water to the extent possible for basic needs for people and the environment.

This chapter assembles the social and environmental issues of water management into a framework to explain how they can be handled in water decisions. Since the Water Resources Planning Act of 1964, these issues have been evaluated extensively, and they were the subject of an Engineering Foundation Conference in the 1980s (Viessman and Schilling, 1986). Reviewing this work of some 25 years ago, it is apparent that social and environmental issues have transcended its focus on water resources projects and reached into micro-level issues such as safe water and local fish and wildlife problems.

ECONOMIC NEEDS FOR WATER

The focus of the chapter is on social and environmental uses of water, but it is also important to explain the category of economic uses. Actually, these go well beyond the obvious categories of water supply for business operations, which require water to operate, and process water for industrial applications such as pulp and paper processing. This type of economic use for water can be traced through economic accounting systems such as the North American Industrial Classification System, which organizes data for many types of industries. Almost any category of industry requires at least some water for economic purposes.

Looking across the water management purposes of supply, wastewater, irrigation, hydropower, recreation and others shows many other economic uses of water. To distinguish them from social and environmental uses is not always easy because you can make the argument that some social and environmental activities also have economic purposes. For example, if water is used to nourish a fishery, is it economic or environmental? Or does it support a local community's social development for fishermen and their families? Fortunately, we do not need to resolve complex questions like these to move ahead with our discussion because so many direct social and environmental uses of water require the attention of governance.

Economic, social and environmental objectives are multi-faceted and inter-related.

SOCIAL NEEDS FOR WATER

Social needs for water refer to any human water uses that are not for economic or environmental purposes. In this broad category, the greatest social need for water is for access to safe water and sanitation. You might say these have economic purposes as well, but the basic need is the social requirement for people to take care of their daily physical requirements. There is broad agreement that it is a human right that these needs should be met, regardless of their ability to pay.

Of course, social needs for water go far beyond these most basic of needs. They reach beyond the point where people would agree that they are actually needs and they begin to become wishes. For example, viewing water scenes as beautiful works of nature is a social use of water but people begin to disagree on who should pay for it. Issues such as this introduce many questions about private property rights, use of resources, and others.

Because social water needs cut a broad swath, their definitions tend to be vague. For example, by its definition TWM "addresses social values" and "fosters public health, safety, and community good will." These abstractions embody important needs and take on different priorities and meanings among different cultures and income groups but they are hard to nail down.

One way to assess social needs for water is to align them with the "hierarchy of human needs" made famous by the psychologist Abraham Maslow (1943). Figure 11.1 illustrates how needs at the base level of this hierarchy include water for survival and physiological needs, whereas more discretionary needs migrate upwards toward water's social value for enjoyment and culture.

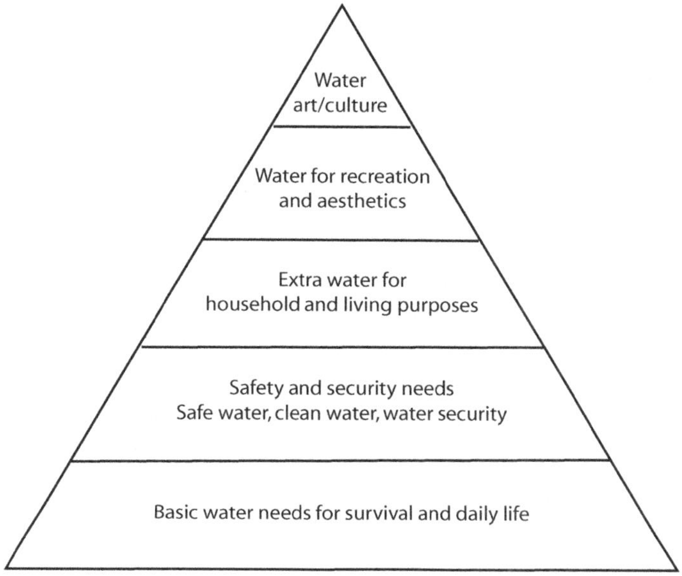

Figure 11.1 A hierarchy of water needs

In between these extremes are found security needs such as public health and freedom from risk of harm, environmental justice, and flood risk. Still higher up, social needs can include convenience issues such as urban drainage and water delivered to home and business, as well as the "good life" with water based recreation and enough water for flowers and lawns.

Social objectives can be framed in a "hierarchy of human needs."

The bottom line of social needs is that the most basic ones are relatively easy to identify, but many others may be significant but much more difficult to measure and plan for.

ENVIRONMENTAL NEEDS

Environmental needs for water are also important and, in many cases, equally difficult to measure and finance. These requirements are exemplified by the concept of the "Tragedy of the Commons," which explains how the environment become degraded due to lack of attention from owners (see Chapter 8).

The need for water to sustain environmental systems is sometimes considered as an issue that applies mostly to developed countries, which are able to fulfill the basic water needs of their populations (UNESCO, 2010). However, when environmental water needs are recognized to include water for fisheries, clean streams, and nourishment of essential plant and animal communities, then they can be viewed more as a basic need everywhere and not only a discretionary use of water for higher-income countries.

Environmental needs are not discretionary—they are basic needs of society and living things.

SOURCES OF WATER FOR ECONOMIC, SOCIAL AND ENVIRONMENTAL PURPOSES

As is the case with other needs, water for social and environmental purposes can be supplied by utility service organizations, through non-utility organizations, or through self-supply. These sources are shown on Figure 11.2, which illustrates the ways that social and environmental water needs are met, along with economic needs, which are met mainly through utility suppliers and through industrial or irrigation self-supply.

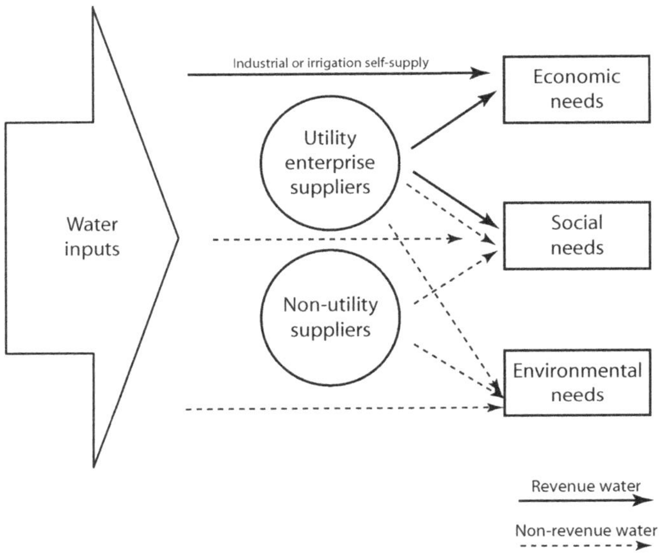

Figure 11.2 Economic, social and environmental uses of water

When utilities provide water for which they are paid, it is called "revenue water." Some of the water they provide for social purposes, such as basic supplies for people who cannot pay the full cost, is in reality "non-revenue" water, and involves cost-shifting, which can be tricky to balance on a fair basis. Utilities in different cultures have varied approaches for dealing with this issue, ranging from simply ignoring the inability of some customers to pay to having formal systems of different rate structures.

While social and environmental needs can be measured by economics, it is difficult to determine how to pay for them.

Utilities also provide some non-revenue water for the environment, as in the case where they are compelled to release flows for instream purposes from their storage reservoirs.

When customers cannot pay for water that utility enterprises provide, the utilities must find other ways to pay for the costs they incur. Sometimes the needs are provided by subsidies from government programs and sometimes by invisible financial transfers within the utilities.

So, provision of water by utilities is like the hard window and soft window of a development bank. The hard window pays the full cost and the soft windows gets subsidies.

Self-supply can be by industries for their economic purposes, for households to meet their social needs, or by the environment, which uses its own water to nourish habitat. The environmental self-supply is like a reserve that must be protected, otherwise all the water might be used for economic and social purposes.

The range of non-utility suppliers might include cooperatives, mutual companies, philanthropic organizations, and various other non-governmental entities. Examples of situations where they provide social or environmental water would include cases where a social agency might pay utility bills for poor customers or an environmental organization might purchase water rights to provide for habitat.

Without rules and enforcement, the incentive structures of utilities or any other enterprises will channel their attention toward customers who can pay. Governance is the mechanism to find the required balance among competing social, environmental and economic interests and to use authority to overcome incentive structures that work against meeting basic needs.

GOVERNANCE FOR SOCIAL USES OF WATER

A starting point for a discussion of social uses of water is the idea that access to water resources should be considered a human right. This idea has to be taken further, however, to explain how much water to provide and the responsibility of users to pay for their water services.

The basic tenets of human rights were included in the 1948 United Nations Universal Resolution on Human Rights. While water is not mentioned specifically, international actions and conferences since 1948 have continuously identified water and sanitation as basic needs of society. These have also recognized environmental needs for water, but social needs receive top billing. So, there is broad international consensus that safe water and sanitation should be available to everyone.

People also require affordable water for other social and economic purposes, such as water-based amenities that lead to better lives. They also need to be protected from hazards that include security against flood damage. Taking these needs together leads to a list of social priorities for water:

- Access to safe drinking water
- Access to basic sanitation services
- Security against flooding
- Environmental justice (protection from environmental hazards)
- Water for social amenities

So, the bottom line of social needs leads to an emphasis on access to safe public services, safety against flood threats, and access to a range of additional water services that includes some obligation to pay for them.

Although access to water is a basic right, allocating water and paying for services can be controversial. For example, if a water utility adds amenities to its water projects, the cost must be absorbed in its rate base and not everyone will agree.

Drawing the line between the public and private good attributes of water is not an exact process. On the side of private goods, you can allocate by setting rates (but not normally through competition) and on the public side, water resources must be allocated by political decision processes. Keep in mind that water resources management refers to all water-related services, including supplies, capacity to assimilate pollutants, environmental nourishment, and security against damage from flooding.

Whereas the environmental impact assessment is used to report total environmental impacts, a social impact assessment can be used to review the social effects of projects and other water management actions. These do not have the legal standing of environmental studies, but they can inventory effects such as public health and safety, equal opportunity, and community good will (International Association of Impact Assessment, 1996; Barrow, 2000).

A listing of the social goals of water resources development shows categories that can be used in impact statements (Technical Committee, 1974):

- Collective security
- Environmental security
- Individual security
- Economic opportunity
- Cultural and community opportunity
- Aesthetic opportunity
- Recreational opportunity
- Individual freedom and variety
- Educational opportunity

In the U.S., social needs for water have received less attention than economic and environmental needs. The Principles and Standards for Planning of Water and Related Land Resources initially listed social well-being as a need, with five categories: real income distribution; life, health, and safety; education, culture, and recreation; emergency preparedness; and other social effects (US Water Resources Council, 1973). However, these categories did not survive to the final versions of the P&S.

Social issues are at the root of issues of poverty and equity in developing countries, which are explained in the next chapter.

GOVERNANCE FOR ENVIRONMENTAL WATER NEEDS

Although environmental water needs may be considered by some as less urgent than basic social needs, to many other people they are considered of very high priority and a matter of survival of the human race. In the U.S. and other higher income countries, environmental needs for water often trump economic needs, such as water for irrigation or industry. The activities of environmental activist groups in opposing water development constitute an important institutional force in water decisions.

U.S. environmental law is discussed in Chapter 14, which presents an overview of the main statutes. These provide for:

- Environmental planning through use of assessments and impact statements
- Water quality control through the Clean Water Act
- Environmental protection through laws about endangered species protection, wild and scenic rivers and habitat
- Control of dredging and filling in streams and waterways
- Licensing of hydroelectric plants
- Control of toxic chemicals
- Global climate change
- Special problems of estuaries and ocean pollution

CONCLUSION

Social and environmental uses of water take place mostly outside the enterprise arena where the cost of service can be paid directly. The starting point to identify the needs for them is at the policy level, but the needs extend to the lowest levels of water operations and should be addressed in all water operations.

Planning and accounting for social and environmental water uses begin with basic issues—such as safe water for all and protection of rivers—but they can quickly become fragmented to deal with many, many complex issues. This should be avoided so as to focus attention on the set of most basic issues and not discourage the public from giving attention to them.

Social and environmental water needs can be met through the collective actions of water organizations, which decide to adopt TBL approaches to water management. There are important roles for the water utility, the regulator, the environmental objector, the water user, or the citizen at large. Stewardship as in TWM requires productive behavior in each role so that the whole system gets better. Management systems should be designed with incentives to encourage people toward behavior that helps everyone.

Postscript

On the first Earth Day in 1970 I was teaching at the University of Denver and prepared a lecture that showed environmental problems that included Cleveland's Cuyahoga River catching fire. It seemed that dealing with environmental issues was an imperative we could not ignore. That same year Congress passed the National Environmental Policy Act and began to require Environmental Impact Statements. It did not take long before projects were choked under mountains of paper and costs that mounted into the $ millions. Government workers, consultants and scientists benefited because the studies created work and contracts for them. I thought, "Surely we can do better than this!"

Since then, I've had many lessons in the problem of reconciling environment and development, but during the 1980s a dramatic case showed me more than others had. It was Colorado's Two Forks Project, which turned out to be a landmark case of conflicts between water utilities and environmentalists. Two Forks was a large water supply project that had been planned for decades by Denver and was vetoed by the Environmental Protection Agency, throwing the water supply planning process into turmoil in spite of an environmental study costing over 40 million in 1980 dollars.

Two Forks is an attractive dam site and had been eyed by Denver's earliest water developers by the 1890's when Denver was served by private water companies. Denver had filed on water rights by 1931 but did not feel a need to act until 1982, when it joined with some 40 suburban governments and water districts to form the Metropolitan Water Providers.

Denver needed a permit under Section 404 of the Clean Water Act (see Chapter 14) and this federal action required an EIS to be prepared by the Corps of Engineers. The Corps issued the EIS in 1988 with a recommendation to issue the permit, but they were overruled by the USEPA.

The conflict lasted the better part of a decade and ended with a vetoed permit application. The main lesson was that getting an environmental permit for new water supplies is an extraordinarily complex task and projects must be planned with care and consider the needs of neighbors and the environment if they are to have any chance to survive all of the guerilla warfare.

The Two Forks conflict included several chapters. It erupted when Denver and suburban water agencies created a joint water project that seemed to be a model for badly-needed regional cooperation. However, after years of negotiation involving two left-leaning governors and much additional political negotiations, environmentalists were able to convince the first Bush Administration to veto the project under Section 404 of the Clean Water Act.

The aftermath of the Two Forks controversy has been dramatic in Colorado. Denver Water changed its orientation from a regional supplier to a hold-the-fort philosophy. Several new organizational initiatives were announced and new consortia have been created. The City of Thornton, which is a northern suburb of Denver, secretly bought up farms and water rights in Northern Colorado, but they have not been able to divert any of the water yet, some twenty years later. Governor Roy Romer (1993), who participated in the process, said we must "blow the whistle on what has become an unacceptable level of administrative gridlock, litigation, expense and delay whenever water development or transfers are proposed." He identified governance concerns that included: waste of public and private funds; environmental consequences, extensive lead time required to produce new supplies; and impact on future development in other parts of the state. He also suggested new governance initiatives such as: a regional water coordinating organization led by the state; state incentives such as loans or grants to encourage cooperation; state water project; cooperation with agricultural water users; and enhanced information and decision support systems by state agencies.

It's amazing to me that when I mention Two Forks to Coloradoans today, few are even aware of the controversy or its lingering after-effects. This lack of awareness, along with the rigidity of institutional constraints that suppress integration and cooperation, suggest that breaking the barriers that prevented a better resolution of Two Forks will be very difficult indeed.

Sometimes it seems like there's nothing new under the sun, but rather old things keep coming up in new forms. The TBL methodology seems like that to those old timers who have been analyzing social and environmental objectives for decades, but it's also true that there is a time for everything, and now is a better time for TBL than in the past due to today's emphasis on sustainability.

It's also true that, if you recognize water's public good attributes, you can see clearly that only looking at economic outcomes is not enough. You must also look at social and environmental outcomes. That's been known since the advent of welfare economics, which considers the maximization of total social welfare, however it is defined.

Sometimes I feel like more of a socialist than a capitalist because TBL makes eminent sense to me. Those of us who subscribe to IWRM implicitly sense that we have to achieve many objectives with water, not just single purpose optimization of a goal like profit or productivity.

Even the TBL display and multi-criteria decision analysis with their multiple objectives do not seem new. However, they were new back in the 1960s when Morris Hill (1968) published a dissertation about a "goals-achievement matrix." Today, the idea of presenting multiple projects and goals in a single display is a common approach.

Although social and environmental goals and MCDA approaches date back in time, their problems have not disappeared. The subjectivity inherent in evaluating non-quantitative goals about which people disagree renders conclusions and business case analysis invalid at times and questionable in many other cases.

We have a long way to go in considering social and environmental objectives fully. The problem is as large as politics itself because so many different value systems are in operation. Still, the TBL approach gives us a way to start talking about finding commons solutions.

REFERENCES

Barrow C. J. (2000). *Social Impact Assessment: An Introduction*. Arnold, London.

Hill M. (1968). Goals achievement matrix for evaluating alternative plans. *J. Am. Inst. Planners*, July, **34**(1), 19–29.

International Association of Impact Assessment (2006). *Principles of Impact Assessment.* http://www.iaia.org (accessed 15 November 2006).

Maslow A. H. (1943). A theory of human motivation. *Psychol. Rev.*, **50**, 370–396.

Romer R. (1993). *The Role for the State of Colorado on Front Range Water Challenges*. Colorado Water Convention, Denver.

Technical Committee of the Water Resources Research Centers of the Thirteen Western States (December 1974). *Water Resources Planning, Social Goals, and Indicators: Methodological Development and Empirical Test.* Utah Water Research Laboratory (PRWG 131-1), Logan, UT.

UNESCO (2010). *Hydrology for the Environment, Life, and Policy.* http://portal.unesco.org (accessed 11 April 2010).

U.S. Water Resources Council (1973). Water and related land resources: principles and standards for planning. Part III. September 10, 1973. *Federal Register*, **38**(174), 24778–24869.

Viessman W. and Schilling K. (eds.) (1986). *Social and Environmental Objectives in Water Resources Planning and Management.* American Society of Civil Engineers, Reston, VA.

Chapter 12

Capacity-building for water management and governance

The term "governance" seems to suggest government control of human activity. Wise political leaders have seen past this limited view, however, and strived to explain the broader purposes of governing. Should government be limited or activist? The debate goes on. Meanwhile, the need for effective governance is more apparent every day in complex spheres of public life, such as management of water.

In particular, the problem of access to water and sanitation in low-income countries requires attention from governance, but it must extend beyond the water sector alone. Pro-poor policies for water and sanitation require attention to political will, broadened scope for governance, including legal and management in utilities and regulators, stakeholder involvement, financial resources, and conflict resolution mechanisms (Osinde, 2008).

One way that governance expands beyond the limited view of government is to recognize that it involves much more than control through laws and regulation. Governance also has a support function that is expressed through the concept of capacity-building. This means that governance helps to provide the means to meet society's needs as well as the control mechanism. Capacity-building may occur through financial assistance, through organizational efforts, through technical assistance or through human resource development.

There is no one formula for capacity-building as part of water governance. This chapter is about the different ways it should be applied across the purposes of water management and the different types of government and social systems.

Capacity-building as a function of governance provides a helping hand for water management.

NEED FOR CAPACITY-BUILDING IN WATER MANAGEMENT

After a massive earthquake struck Haiti in January 2010, access to safe water became an immediate recovery need. While this is frequently the case during emergencies, billions of people around the world also lack access

 Governance and Management for Sustainable Water Systems. By Neil S Grigg. ISBN: 9781843393467. Published by IWA Publishing, London, UK.

to safe water and sanitation on a daily basis leading to chronic health problems. Hunger stalks people in many places due to lack of irrigation water, and people who bathe and wash clothes in polluted water become more vulnerable to flood hazard.

Hunger is a serious and chronic water-related problem too. Writing about the one billion plus hungry people in the world, Geithner and Gates (2010) explained: "They are also the ones who face shortages when rainfall patterns change and reduce the amount of available water."

Countries at all levels face similar issues, and they align themselves along a spectrum that varies with income, government capacity, and social cohesion. While water issues seem most urgent in developing countries, billions more in a middle zone have water services that are less safe, reliable, or affordable than they should be.

Figure 12.1 shows four cases of water organizations in varying situations, from large utilities in high-income countries to small service providers in low-income countries. This wide range of organizations will require different approaches to empowerment and regulation. The large, high-income cases will be much more able to take care of their own empowerment needs, and they will require different approaches to regulation. The smaller, low-income cases will require much more empowerment than regulation, so that they can improve their capacities and deliver better services.

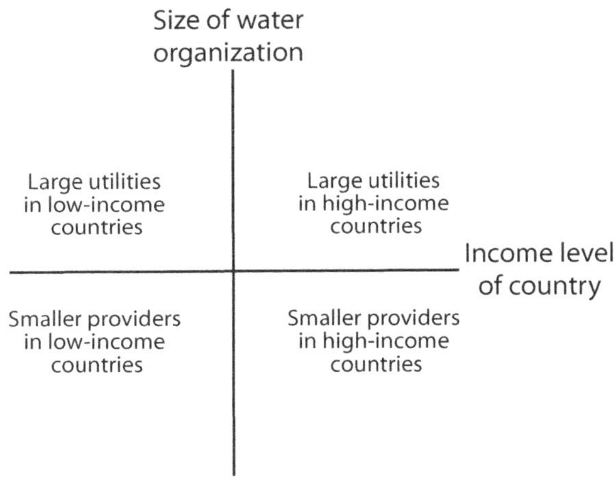

Figure 12.1 Water governance scenarios

This chapter examines how capacity-building as a task of governance should be carried out. By looking at the inputs required for adequate water services, we can identify the general elements of capacity-building required and how these vary from situation to situation. The chapter explains these elements, why they are required, and how governance should be structured to provide them.

The discussion in the chapter goes to the larger question of the limits of government responsibility to meet social needs. It is part of the debate over the merits of capitalism versus socialism, and there is no single correct answer to the question. To justify the proposition that it is a proper role of governance to address basic social needs as well as to regulate, the reader is referred back to the broader definition of governance presented in Chapter 1, which was that governance goes beyond the control of private activity and addresses the broad needs of society.

FROM CAPACITY-BUILDING TO REGULATION

The model of water services presented in Chapter 2 outlined a spectrum that ranges from the most essential (water supply) to more discretionary services. However, all of the services are needed everywhere, it is just a matter of the degree and timing of their provision. For example, stormwater service in an upscale neighborhood in Europe will be much different from basic drainage in a low-income neighborhood of a developing country, but both require drainage for essential as well as discretionary purposes.

This section explores the ranges of water services from the low to the high end of service quality. At the high end, less capacity-building is required, but at the low end, the systems may need life support if they cannot be strengthened. The highest quality services can in some cases serve as benchmarks, but these are not attainable everywhere due to income limitations.

Table 12.1 Capacity-building and organizational actions

	Highest quality services	**Lowest quality services**
Water supply	Practically universal access to reliable and safe water. Effective regulation of water withdrawals and drinking water quality.	Access to safe water systems very limited or non-existent. Ineffective or missing regulatory systems.
Wastewater	Universal access to sanitation. Effective wastewater treatment and water quality regulation.	Limited access to sanitation. Lack of basic wastewater treatment. No water quality regulation.
Irrigation and farm drainage	Infrastructure adequate. Systems in place to mediate conflicts over supply, pollution, and policy. Some water pricing.	Poor infrastructure. Lack of management and mediating institutions. Systems require much assistance.
Instream flows	Effective mediation processes for instream flow conflicts.	No systems to mediate instream conflicts.
Stormwater and flood management	Local to national stormwater programs in place. Effective structural and non-structural flood programs.	Ineffective or missing stormwater systems. Many flood hazards, especially illegal residents along streams.
Areawide management	Effective agencies exist to manage and resolve conflicts.	Ineffective or missing areawide agencies.
Regulatory programs	Effective regulatory programs for water allocation, public supply and discharges.	Ineffective or missing regulatory programs.

By examining the table above (Table 12.1), we can identify how the need for capacity-building spans the range of organizational actions from the simple case of village level to the complex case of large urban areas and regions.

We add to this identification of needs a discussion of the general elements of capacity, which are an effective institutional framework, resource inputs, infrastructure, organizational capacity, and human resources capacity (Global Development Research Center, 2007). There is a strong correlation between financial capacity and availability of these inter-related inputs. Capacity gaps can occur in low-income countries or low-income regions of a high-income country. For example, in the U.S., many of the smallest water utilities are in rural areas with limited financial capacity.

The following table (Table 12.2), shows how capacity needs align with water governance issues from Table 12.1.

Table 12.2 Capacity needs and governance

Issue	Capacity needs
Access to safe water systems very limited or non-existent	The response in low-income situations depends on the regulatory requires and vary from a developed country model, where subsidies may be required, to a less-developed country scenario where appropriate but unregulated systems can be provided by donors.
Limited access to sanitation. Basic wastewater treatment not in place.	Similar to the safe water case, access to sanitation can vary from the highly-regulated scenario where uniform rules may mandate subsides, to less-developed country scenarios where appropriate but unregulated systems are effective. Wastewater treatment is expensive and both capital and know-how are needed for appropriate systems
Poor I&D infrastructure. Lack of management and mediating institutions for irrigation.	In high-income countries, irrigators may pay for their infrastructure and operating costs and mediating institutions are provided. In low-income scenarios, ineffective systems may be closely linked to rural development policy and require high levels intervention by government and donors.
Ineffective or missing stormwater systems. Many flood problems, especially with illegal residents along streams.	Given the local character of stormwater systems, if residents are informed, they can handle many problems themselves. However, in large urban areas of any country they cannot and government intervention is required. The problem of flood hazard to illegal residents is one of the most intractable water issues and cannot be solved by water governance measures alone.
Ineffective or missing areawide agencies.	Where governance capacity is high, local residents will normally identify the potential to organize areawide agencies and take action. In low-capacity cases, action is needed by external forces, either the government or international agencies and donors.
Ineffective or missing regulatory programs.	For high-income countries, ineffective or missing regulatory programs call for government reform. In lower-income cases, it is a matter of appropriate regulation and to select the mixture of empowerment, coaching and nurturing, and regulatory controls.

CAPACITY-BUILDING FOR WATER SERVICES

Capacity-building relates to empowerment and is the vehicle to enable water organizations to function effectively to provide the required services and controls. The term capacity explains important inputs to total operational effectiveness, such as those shown on Figure 12.2. The figure shows institutional, financial, technical, and workforce capacity, both for the water management organization itself and the external environment, which includes institutions such as those explained in Chapter 3 that included regulators, the judiciary, government agencies, and many others.

Capacity can be modeled in different ways, and Figure 12.2 divides institutional capacity into the internal (water management organization) aspects and the external (all other institutional arrangements). Then, financial, technical and workforce capacity are shown as applying to both the internal and external cases. For example, if the management organization had a strong technical capacity but the regulators lacked one, governance would be out of balance.

Institutional capacity involves both the external and internal operating environments. If government bureaucracy is too unworkable, it will hinder water management. On the other hand, organizations themselves can be guilty of tolerating poor institutional environments. Institutional capacity deals with controls, rules, regulations, and requirements. Rules should clear the way for effective and responsible action but not put up unneeded barriers.

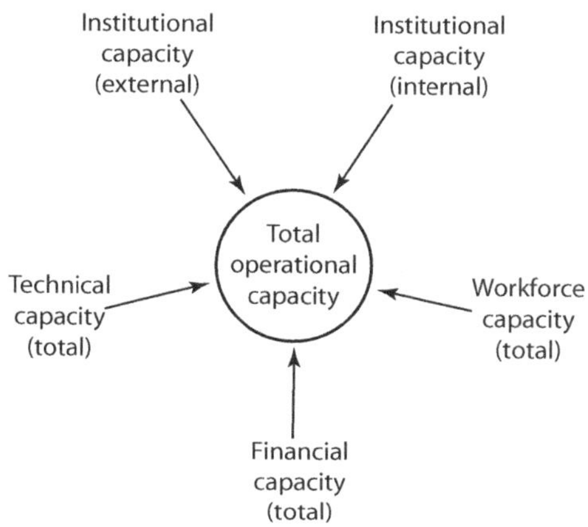

Figure 12.2 Inputs to total operational capacity

Water supply empowerment

While many believe that lack of safe water is the result of ignorance and that the solution is simply to drill wells for low-income people, the reality is much more complex and illustrates the need for capacity-building. It begins with the recognition of need, which is part of a bottom-up process that involves planning, financing, designing, and constructing, and above all, maintaining. If local people can do these themselves, they normally will. If they lack knowledge, capacity, and resources to establish and maintain service, they will require external assistance.

The required functions illustrate how more than construction is required and the needed empowerment and regulatory structure will vary with sophistication and scale of services. Even in the need-recognition stage, no one may recognize the need and intervention might be needed. It is best to establish service on a bottom-up basis, and community leadership is required for local initiative. No on may know how to organize service, so intervention in the form of public or private initiatives might be required. If the community lacks resources (money, knowhow, equipment, energy, infrastructure) these might need to be provided through a grant or loan.

The organization phase depends on how the services are established, and organizational knowhow is required. In the same sense, to have good O&M and organizational management requires knowhow.

Regulatory systems require a willingness to comply and establishment of a community control system that people buy into.

Users are involved as well, and they require education and training to use systems properly.

To summarize this discussion, the following empowerment elements may be needed:

- Intervention to establish the need (motivator)
- Leadership to spark the service (leadership)
- Public or private initiatives for establishment of service (incentives)
- Organizational institution (management)
- Resources (money, equipment, energy, water, infrastructure)
- Organizational and management knowhow (knowhow)

- Willingness to comply (culture)
- Establishment of a community control system (governance institution)

So how does governance go about getting these done? The answer depends on the context. An external motivator might intervene to establish the need and this might require policy studies and local discussions. To spark the service may require leadership to move things along and overcome obstacles. Public or private initiatives will depend on which is appropriate. In some cases, a community organization might be appropriate but in others a private scheme might do better. In either case an organizational and management institution will be required, and resources in the form of money, equipment, energy, water, and infrastructure must be mobilized. Willingness to comply is a cultural issue and some kind of plan to gain buy-in will be required. Establishment of a community control system as a governance institution will also depend on the context, and it could vary from something as simple as independent checks of system safety to a formal regulatory structure.

As an example of capacity-building for water supply, Hukka (1998) examined in a doctoral thesis the capacity development for utilities in places such as Kenya. He concluded that financial capacity was essential and, using data from around the world, he developed a three-tiered model for governance. It showed roles for the central government (promulgate institutional framework); the basin organization (policy and monitoring); and the local service provider (manage core competencies for production and service delivery). He also identified roles for donors and private entrepreneurs.

Sanitation and wastewater treatment

While providing access to sanitation and wastewater services seems similar to water supply, there are important differences. Institutional development will present a different challenge for wastewater than for water supply.

If access to sanitation service is limited, people will take it on themselves to find some solution, and their actions have strong cultural roots that should be respected. However, to have safe, convenient and healthy conditions may require similar organizational actions as local water supply. In particular, finding a way to pay for the service may present more of a challenge in sanitation than in water supply.

Providing wastewater services might be a stepwise process. Sewer hookups to convey waste from buildings provide convenience and might be a popular move, but to collect funds to construct and operate wastewater treatment systems is a different matter. If basic wastewater treatment is not in place, then the challenge of the public coming to understand the need for it must be confronted. Then, it is necessary to build the institutional structure to make it happen. Wastewater treatment is expensive and both capital and know-how are needed for appropriate systems.

Irrigation and drainage

While irrigation and drainage services seem similar to water and wastewater in configuration, there are important differences, particularly in how water users participate. With irrigation systems, management and mediating institutions deal with farmers rather than water supply customers.

To visualize the need for capacity-building in irrigation, think about how the farmer is in touch with the needs of his crop and is fully engaged in bringing water and other inputs at needed times to optimize his production and income. He must not only have the knowhow to farm, he must be able and willing to cooperate with others to manage the irrigation system.

Even at small scales, the tail-ender problem can occur where upper irrigators take all of the water, thus leaving none for the lower irrigators. This is a similar problem to the one of coordinating all water withdrawals along a stream (see Chapter 4), but in this case the irrigators must have an organization to coordinate among themselves. The need for this explains the emergence of the concept of mutual irrigation organizations, which exist around the world to provide management and coordination. Empowerment is needed to establish and nurture these organizations.

The concept of the agricultural extension service that is well-established in the U.S. and also used in other countries illustrates the need for capacity-building in irrigation as part of the overall farm system. Extension agents should be knowledgeable about farming practices across-the-board and specialists are required for irrigation practices. In addition, financial capacity is required through the agricultural finance system, including access to credit.

Stormwater and flood systems

Stormwater and flood systems present unique institutional problems in the sense that you may have ineffective or missing stormwater systems and many flood problems are exacerbated by illegal residents along streams. To see the consequences of these, imagine a large city immobilized by local flooding even from a small storm or the misery that accompanies large scale flooding through a poor community along a river.

These water problems are different from those of local water and wastewater systems in that they involve drainage basins ranging from small to large and capacity-building requires coordinating organizations.

Akmalah (2009) used systems analysis tools to study flooding in Jakarta, and explained how it is a complex socio-technical problem that requires an integrated solution. Her analysis showed the links between attractiveness of the city, migration, poverty, lack of community cohesion, overwhelmed infrastructure and management systems, and the resulting flood disasters. The required solution would feature community participation and a combination of structural and non-structural measures. The institutional framework would have to ensure a commitment to coordination, resources mobilization and enhancement of preparedness.

Katsuhama (2010) found that developing countries must improve institutional, organizational, and individual capacities to address flood threats from climate change. His research developed a tool to evaluate how capacity affects the flood management systems and to identify the needs for capacity building. While institutional arrangements are essential, their effectiveness depends on the capacity to implement them. The limitations of flood control infrastructure must be recognized and should not induce complacency and awareness of flood threats by the local community and data accessibility are critical requisites to fight floods. His case studies suggested that problems and self-sufficiency differ between urban and rural areas. In any case, capacity-building was considered as the fastest and most effective path to effective flood management under climate change.

Areawide agencies

The need for capacity-building in water services illustrates how areawide agencies such as those explained in Chapter 10 can be used to bridge gaps. Ineffective or missing areawide agencies present unique problems, which vary from place to place. As examples, regional water and wastewater services might supplant local systems and create efficiencies that overcome resource problems. The coordination needed to fight floods calls for areawide solutions. Areawide solutions require political vision and cooperation, which can call for leadership capacity as a starting point to initiate solutions.

Regulatory programs

Ineffective or missing regulatory programs (withdrawals, environmental water quality, water and wastewater quality, instream flows, groundwater, non-point sources) illustrate the need for institutional capacity-building to change the context to one where law enforcement is effective.

CAPACITY FOR WATER GOVERNANCE IN DEVELOPING COUNTRIES

Water governance takes on special urgency in developing countries where many people lack even the most basic services such as safe water and sanitation. After these are provided the countries can pay attention to broader water needs, although issues such as water for wildlife and subsistence farming may also be closely linked to survival and need attention urgently.

While countries face different situations, the people most at risk live where income levels and governmental capacities are lowest. However, it is important to recognize that people at a range of income levels require the same kind of empowerment as those in the lowest-income countries. The management of a small water utility in the United States may require capacity-building just as they do in lower-income nations. This occurs because, in all countries, the distribution of wealth inevitably creates disparities in public services.

Given the low capacity for governance in many developing countries, empowerment is needed to lift them from poverty and hardship and to address water management from basic services to broader needs. This is a vexing problem and the main barriers involve low institutional capacity to manage water and lack of access to financial resources.

It is difficult to classify the need for empowerment because of the disparities of wealth inside of countries, but Figure 12.3 makes an attempt to explain how income levels and institutional capacity correlate with water services. If you plot the status of each of the some 180 countries in the world they would indicate at the lower end those with the most deprivation and at the upper end the advanced and wealthier countries. The term "low-capacity water institutions" carries the message that water services are inadequate, but the spectrum is continuous because all water institutions require continuous improvement, even in high-income countries. While a country may graduate to the next level, it never escapes the need for improvement.

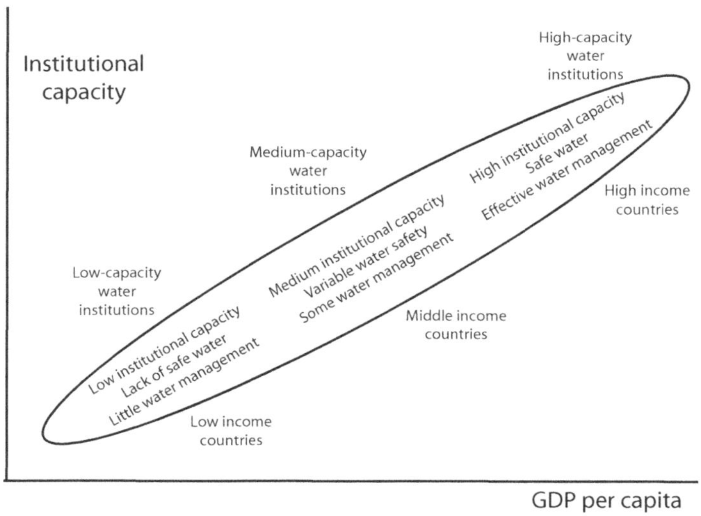

Figure 12.3 Water governance and development level

The symptom of poverty in developing countries is evident in rural and urban areas alike. Around the world, urban areas contain shantytowns with many different names, such as squatter areas, slums, informal or illegal settlements, favelas (Brazil), and bustees (India). In these areas low public health levels would be expected. Their water problems seem daunting, at the same time that they face many other issues. Rural areas contain much poverty as well, often masked by remote locations and lower population densities.

While a country may build new water infrastructure systems, legacy infrastructure systems may be in poor condition and cause performance problems. Urbanization may outstrip capacity to build, renew and manage systems. Water is managed best at the local level, and local governments may not be empowered.

Problems the countries must confront include income levels too low to support developed country service models, low financial capacity, weak education systems, and workforce capacity not up to operating complex management and technical systems. Community coherence may be weak due to conflicts and other causes, and cultures may work against planning, cooperation and management systems. In some countries, the cultural systems require women to carry water.

The water management improvements where governance can make a difference will normally include: improve access to safe water systems and sanitation, to implement wastewater management and fight stream pollution, to strengthen irrigation and drainage systems, to implement stormwater management and reduce vulnerability in floodplains, to increase management capacity of water institutions, and to introduce appropriate regulatory systems.

The water and sanitation problems of developing countries have been on the radar screen for a long time as studies from the 1970s showed that a billion or more people lacked access to safe drinking water and/or sanitation, and suffer disease and infant mortality (Pacey, 1977). The United Nations proclaimed the 1980's as the "International Drinking Water Supply and Sanitation Decade" (Canemark, 1989). Okun (1991) reported that in his 50 years of work, water and sanitation problems in urban areas had become worse and he emphasized the need for capacity-building for a favorable policy climate, appropriate institutional development, and sound management systems.

The Global Consultation on Safe Water and Sanitation for the 1990's emphasized "some for all" and identified the needs for integrated management of water and wastes; institutional reforms to change procedures, programs, attitudes, and behavior and to promote participation of women; community management of services and stronger local institutions; and sound financial practices with appropriate technologies.

Triweko (1992) attributed the low service levels in urban areas to inability to improve and maintain the continuity of service. He also emphasized the need for capacity-building for political environments, appropriate technologies, institutional development, human resources development, community management, and strategies for service improvement.

Lessons learned from the US Agency for International Development's (1990) Water and Sanitation for Health (WASH) project show many needs for empowerment. They found that technical assistance is most successful when it helps people learn to do things for themselves, that the measure of success is the ability to perform effectively after donor assistance has been terminated, and that key participants must shoulders their responsibilities.

Capacity-building is needed for broad participation, not just by government; establishing autonomy in finances, personnel, policies and planning; providing effective leaders and skilled managers; hiring, training, and retaining personnel; setting up administrative procedures and policies; maintaining a commercial orientation and financially sound practices; making sound technical decisions in all areas; creating a positive organizational culture; and providing quality services to consumers (Interagency Task Force, 1992).

Current situation

These lessons can be mapped onto the current situation, which shows improvements but still daunting problems. The 2010 UN-Water Global Annual Assessment of Sanitation and Drinking-Water (GLAAS) report showed a drop in donor assistance to solve the problem (World Health Organization, 2010a). This happened in spite of evidence that improved water and sanitation would lower health-care costs, increase school attendance and boost productivity. UN-Water and the World Health Organization (WHO) estimated that improved access to sanitation and water produces impressive economic benefits and increases country's gross domestic product significantly. One estimate was that poor water, sanitation and hygiene services claim the lives of some 1.5 million children under the age of 5 every year from diarrhoea alone, a greater toll than the combined impact of HIV and AIDS, malaria, and tuberculosis.

Access to sanitation remains a critical issue in developing countries. WHO (2010b) reported that around 2.6 billion people still lack access to adequate sanitation, with most concentrated in sub-Saharan Africa, southern Asia and eastern Asia. The public health consequences are dire: people defecate in exposed areas, thus increasing risk of diseases such as diarrhea, cholera, dysentery, typhoid, and hepatitis A. Sanitation is key in emergency response to disasters, which affect around 200 million people every year. Adequate sanitation helps overcoming poverty by keeping children in school, particularly girls. The economic benefits of sanitation are high and they impact poor children and disadvantaged communities the most.

CONCLUSION

Governance of water services is critical in some developing countries where basic needs are not being met for everyone. It is difficult to generalize about the situations across the world, but the barriers to solving water problems almost always involve low institutional capacity and lack of access to financial resources.

Low institutional capacity is directly linked to the situation of "water governance in a state of confusion" (World Water Assessment Programme, 2006). Issues behind the confusion are lack or fragmentation of water institutions, conflicting decision making structures and upstream and downstream interests, and downright corruption. These problems exist in all countries, but those with low institutional capacity require assistance to develop established institutions and well-organized water services.

Where institutional capacity is low, national or multinational water companies may find opportunities to introduce business services. In some cases these can be beneficial, but in all cases, capacity for local solutions must be developed.

The needs for capacity-building cut across the water purposes, including: access to safe water and sanitation; reducing stream pollution; strengthening irrigation and drainage; utilizing water for economic purposes; protecting groundwater; implementing urban stormwater systems; and reducing vulnerability in floodplains.

Governance faces many challenges, but the way forward will involve shared policy–making, appropriate regulatory systems, and a focus on empowerment to increase institutional, management, and financial capacity.

In a developed country, you would expect policy to be made entirely in-country and at all three levels of government. In a developing country with low government capacity, policy-making might be a shared activity that includes partners, sponsors, and donors, as well as the country's government officials. In extreme cases, donors may not rely on government at all and work directly with local people and communities.

Empowerment in developing countries is a much more important issue than in countries with high levels of institutional, financial, and technical capacity and has been a major focus of foreign assistance programs for decades.

If basic systems to manage water are not in place, it is not appropriate to impose stringent regulatory controls that no one can meet. These controls may be "on the books," but they may not be operational. On the other hand, planning and coordination may offer more important controls on water management than regulations do.

Postscript

The main ideas for this chapter came from my experiences in training and education that reach back many years. In the Army I learned that training and building the capacity of soldiers and units was the key to strength. Military capacity-building also includes equipping and supplying the troops. Perhaps the most important ingredient is leadership to motivate them to sustain hard efforts in difficult conditions.

After I left the Army and entered graduate school and then business I continued to see the need for capacity-building, but it was only after working on problems of small water systems and in developing countries that the range of needs started to become apparent. I credit especially my colleagues at Colorado State University with showing me the importance of university outreach and extension in equipping people and organizations. In particular, I gained an appreciation for what is called the Land Grant Mission of selected universities. These were established under the 1862 Land Grant Act, which provided land to universities in each state for the purpose of establishing schools for teaching, research and extension. At Colorado State, extension focused on irrigated farming and solving problems of water management. One of the first professors of irrigation was Elwood Mead, who later wrote the water code for Wyoming, became Commissioner of Reclamation, and for whom Lake Mead above Hoover Dam is named.

These university extension programs have helped many developing countries. In Chapter 7 I mentioned the large improvements wrought by university scientists in Egypt and Pakistan. Perhaps the leading example of such extension is the work of Nobel Laureate Norman Borlaug in stimulating the "Green Revolution." Rather than at a university he worked mainly through research institutes and with philanthropic funding, but the principles are the same.

Capacity-building is a complex notion. As you read in the chapter above, we view it in the institutional, financial, technical and human resources categories. These are needed in varied doses for different situations. The need to introduce modern management methods in an urban utility is much different than straightening out the operation of a small irrigation company in a rural area. Providing training for operators builds workforce capacity, but it also strengthens institutions and adds technical capacity at the same time.

One experience I benefited from was to assist UNESCO in its project to establish graduate water education at the Institute for Applied Hydrology (IPH) at the Federal University of Rio Grande do Sul, Brazil. The project provided graduate training for the faculty, and several of them went on to become national and international leaders in water resources education and management. Today, IPH is a national center of excellence for education and research for water resources in Brazil.

You also learn from failures. I had an experience in a developing country where my job was to help them develop a national water planning program. We interviewed agencies and professionals and wrote our expatriate report. My team leader, who was much more experienced than I was, suggested a revision to my report because I didn't need to lecture them on how to do water planning. That was a valuable suggestion to me, so I revised the report, but our mission came to naught anyway because that country entered a negative tailspin into civil disorder and chaos, so government systems did not work in any case.

Water governance includes many elements, to be applied in many different situations. Many times, the most important elements of it will be to build capacity.

REFERENCES

Akmalah E. (2009). *Integrated Flood Management Model: A Socio-Technical Systems Approach to Overcome Institutional Problems in Jakarta.* Ph.D. Dissertation, Colorado State University, Fort Collins, Colorado.

Canemark C. (November 14, 1989). *The Decade and After: Lessons from the 80s for the 90s and Beyond.* World Water 89, London.

Geithner T. and Gates B. (2010). A new initiative to feed the world. *Wall Street J.* **CCLV**(93), A22.

Global Development Research Center (2007). *Capacity-Building.* http://www.gdrc.org/ (accessed 6 October 2007).

Hukka J. J. (1998). *Institutions, Organizations and Viable Water Services: A Capacity Development Model for Drinking Water Provision and Production.* Thesis for Doctor of Technology, Publication 230, Tampere University of Technology, Tampere, Finland.

Katsuhama Y. (2010). *Capacity Building for Flood Management in Developing Countries Under Climate Change.* Ph.D. Dissertation, Colorado State University, Fort Collins, Colorado.

Okun D. A. (May 1991). Meeting the need for water and sanitation for urban populations. In: *The Abel Wolman Distinguished Lecture.* National Research Council, Washington, DC.

Osinde R. (2008). *Review of Existing Concepts of Water Governance and an Analysis of Pro-Poor Approaches to Un-Habitat Interventions.* United Nations Human Settlements Programme, Nairobi.

Pacey A. (ed.) (1977). *Water for the Thousand Millions.* Pergamon Press, Oxford.

Triweko R. (1992). *A Paradigm of Water Supply Management in Urban Areas of Developing Countries.* Ph.D. Dissertation. Colorado State University, Fall, Fort Collins, Colorado.

U.S. Agency for International Development (1990). *Lessons Learned from the WASH Project.* USAID, Water and Sanitation for Health Project, Washington, DC.

World Health Organization (2010a). *UN-Water Global Annual Assessment of Sanitation and Drinking-Water (GLAAS).* http://www.who.int/water_sanitation_health/glaas (accessed 21 June 2010).

World Health Organization (2010b). *WHO 10 Facts on Sanitation.* http://www.who.int/features/factfiles/sanitation/facts/en/index8.html (accessed 21 June 2010).

World Water Assessment Programme (2006). *Water: A Shared Responsibility.* The United Nations World Water Development Report 2. UNESCO, Paris and Berghahn Books, New York.

Chapter 13

Transboundary water governance

Water governance is a top-down and bottom-up process. At the top, we need water policies, programs and controls that set the general directions for the sectors of water management. At the bottom, we need effective governing boards and local controls that work for everyone. When you start with a particular issue, like siting a wastewater treatment plant for example, the governance issues may seem complex, but they are relatively simple compared to the multi-purpose, inter-jurisdictional problems you have in a watershed or river basin. So water governance becomes more complex as the scenarios are scaled up to include more players.

As long as you are inside of a single jurisdiction, responsibility for final decision making is mostly focused. It will be difficult to reach consensus, but that is what the political process is for. Once you get involved in a transboundary issue where two or more sovereign authorities are involved, decision making becomes an order-of-magnitude more complex.

For example, in the U.S., water inside a state is subject to state decision making, with some exceptions on matters of national interest. When two states disagree, the Supreme Court is the final authority, unless all parties can agree on a Congressional solution, such as an interstate compact. At the state level, the state governments are the sovereigns, at least over matters they control. Once the issue rises to the national level, the federal government is the sovereign. Between two or more countries where there is no sovereign above them to resolve disputes, how do you deal with the issues? These are some of the issues that arise in transboundary water management.

Transboundary water issues are complex to resolve because of the lack of single decision making authorities.

THE CHALLENGE OF TRANSBOUNDARY WATER GOVERNANCE

Transboundary water issues occur when water moves across political jurisdictions beyond the reach of a single authority that is empowered to work out disputes and balance the competing claims on water resources. So, in addition to the usual water conflicts among stakeholders that you have in any jurisdiction, the inter-jurisdictional conflicts must be worked out. Coordinating water decisions is difficult enough when a single jurisdiction, such as a state or county government has control, but when streams or aquifers cross political boundaries, the decision authority moves beyond a single focal point of decision making to a situation with diffused authority, and the problems become more difficult.

This distinction between single jurisdictions and inter-jurisdictional settings was explained by (Ganoulis, *et al.*, 1996) this way. Internal water management issues include disparities between regions, fluctuations in time scales, inequality between needs and supply, conflicts among users, and institutional, legal, economic, and social issues. Once the situation becomes a transboundary issue, the differences between institutional structures, economic instruments, and international relations enter the picture.

Transboundary water issues are hard to define because so many different cases occur. A basic definition is that transboundary water issues occur whenever water moves across political or river basin boundaries. Almost every stream and aquifer crosses some political jurisdictions, but crossing river basin boundaries only occurs with interbasin transfer. So, when the water crosses jurisdictions we only have a transboundary issue if the jurisdictions are those that manage the water. For example, water crossing from one U.S. state to another creates an interstate water issue. Water crossing over a county line does not really create an issue because the counties are not the major management entities, although issues of concern can arise.

A transboundary water issue occurs whenever water moves across political or river basin boundaries.

Many transboundary issues arise with cases that involve: water allocation, water quality, lakes, aquifers, flooding, and instream flow issues such as environmental water. These can occur at scales from small watersheds up to large international river basins. These many scenarios make it difficult to classify transboundary issues, but a proposed system to do that is offered in the next section.

This chapter explains the special governance requirements of transboundary water issues. Its topics are closely related to those in Chapter 10 about integration through multipurpose, river basin, and regional organizations. This chapter deals specifically with the management of conflicts that arise in transboundary situations.

CLASSIFICATION OF TRANSBOUNDARY WATER ISSUES

The proposed classification system for transboundary water issues illustrates the range of choices. The variables in the classification framework are type of water management purpose, whether surface or groundwater, water quality or quantity, and the political scale involved. This yields a set of lists to use to classify scenarios.

The classification system is illustrated in Figure 13.1, which provides a menu of transboundary issues. You start by selecting the geo-political scale and identifying the issues as water rights, water quality, flooding, or other issues. Then, you determine if surface or groundwater is involved (or both) and what the demands for water use are.

The four variables shown in Figure 13.1 enable us to assemble scenarios such as:

- Surface or ground water allocation across borders (such as an interstate compact)
- Interbasin transfers (basins-of-origin lose water to importing basins)

- Water quality issues across borders (such as upstream-downstream pollution)
- A river that forms a boundary (such as the Rhine)
- A lake that straddles borders (such as Lake Constance in Europe)
- Cross-border aquifers (such as Nubian Sandstone of Egypt, Libya, Chad, Sudan)
- Complex issues crossing multiple borders (such as the Danube)

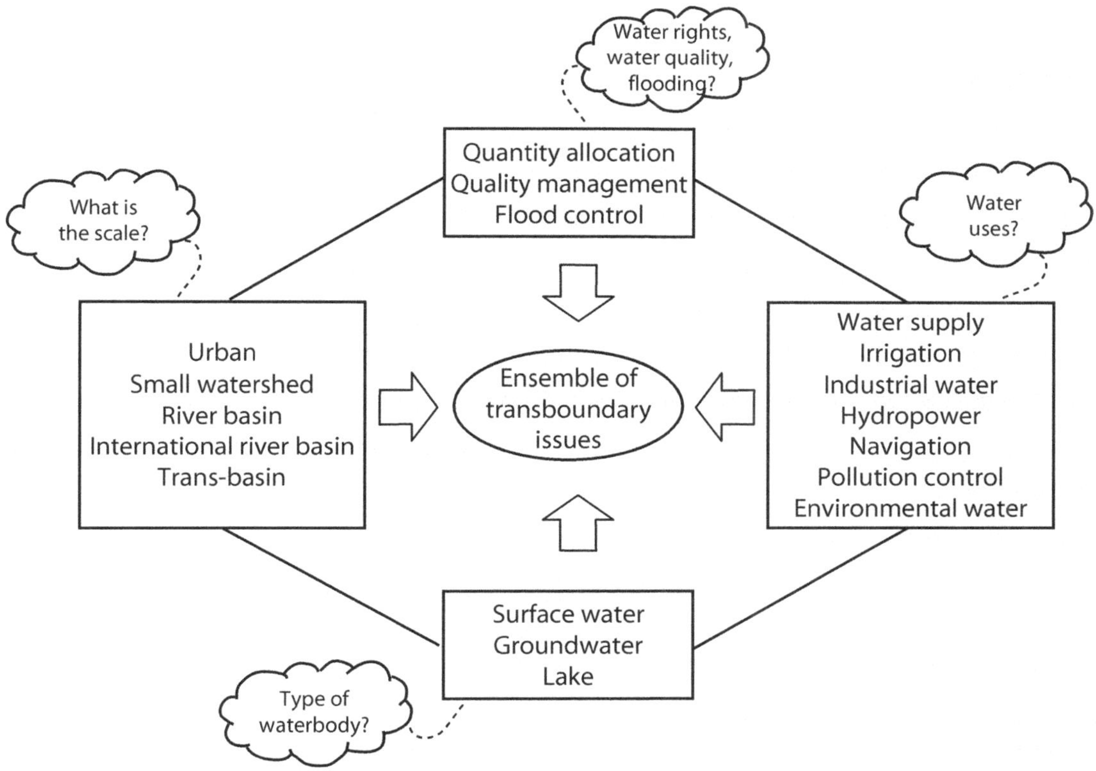

Figure 13.1 Types of transboundary issues

DATABASES ON INTERNATIONAL RIVERS

Transboundary river issues and agreements are studied by a number of international institutions and research institutes. Two programs with significant activities are in the World Bank and at Oregon State University.

World Bank

The World Bank (2009) is involved in management of international waters. It reports that the world's more than 260 international rivers involving countries with about 90 percent of global population can create tension with consequences that reach beyond the rivers. If there is no overall authority or legal mechanism, management instruments to facilitate regional cooperation and effective institutions and investment are needed.

Transboundary water issues addressed in World Bank operations include water allocation and sharing, pollution control and water quality management, and protection of freshwater, coastal, and marine ecosystems. The Bank works with the Global Environment Facility (GEF) to support regional water and environment programs for management of shared waters. Examples include the Baltic Sea, Black Sea, Danube River Basin, and shared lakes such as Lake Ohrid and Lake Victoria. This role of the Bank leads to its roles as a mediator to confront sensitive transboundary issues, such as on the Nile Basin Initiative (NBI). The fact that the Bank might bring resources to the table offers an empowerment facet to this work.

Oregon State University

A research center led by Professor Aaron Wolf at Oregon State University (2009) maintains a set of databases on transboundary freshwater disputes (Transboundary Freshwater Dispute Database). The databases contain a large amount of information in collections that include: a Middle East Water Collection; Case Studies on Water Conflict Resolution; an International Freshwater Treaties Database (some 450 agreements made from 1820 to 2007); an International Water Event Database (historical water relations from 1948 to 2005); an International River Basin Register; a U.S Interstate Freshwater Compacts Database (39 US interstate compacts); and a Water Conflict and Cooperation Bibliography.

GOVERNANCE OF TRANSBOUNDARY WATER ISSUES

When water decisions reach beyond a single authority, inter-governmental agreements are needed to coordinate among the competing interests across the political jurisdictions. This creates a need for two levels of governance: within the political entities and at the level of shared governance.

Shared governance is a complex and visionary concept but it is needed at all governmental levels when inter-jurisdictional disputes arise. Examples range from low level issues such as if two small cities share a common swimming pool to global issues such as nuclear disarmament. It means a sharing of power to make decisions and it starts with a shared vision to create outcomes that work for all parties. The shared vision is based on the perception of a win-win outcome from reaching an agreement.

Shared governance and shared visions are needed for effective resolution of transboundary water issues.

The elements of shared governance for transboundary water issues are: policy to deal with negotiations among multiple authorities; empowerment of the coordination mechanism to balance water–sharing; formal inter-jurisdictional agreements; and appropriate controls based on planning and coordination and with regulation for compliance with the agreements.

Many transboundary negotiations are bilateral, but some international laws and regional agreements exist. An example is the regional convention for the Protection and Use of Transboundary Watercourses and International Lakes (Water Convention), established in 1992 by the UN Economic Commission for Europe. Also, the UN Convention on the Law of the Non-Navigational Uses of International Watercourses is an international legal framework for guiding transboundary water conflicts. It consists of guidelines without sanctions, and has been ratified by sixteen countries, but is not fully operational.

Muys, Sherk, and O'Leary (2007) developed a model framework for a U.S. interstate water compact, and with appropriate modifications its provisions could apply as well to international agreements. It includes seven articles for: purposes and general issues; timelines; definitions; the coordination mechanism (a commission); water apportionments based on safe annual yield; water quality; and details of the management program.

EXAMPLES OF TRANSBOUNDARY SCENARIOS

In this section a few examples of scenarios such as those listed in the classification system given above are summarized.

Water allocation across borders

Water allocation is probably the most contentious transboundary issue because it has similarities to struggles over property rights and carries rights to water use and the benefits that come with them. In some legal systems water rights are property rights, and even a long term license may seem like a property right.

In the U.S. system, water allocation across boundaries is normally done by interstate compacts. More than 200 compacts exist for interstate waters, interstate transportation, and law enforcement. Of these, at least 76 address natural resource issues, such as conservation and environment, energy, planning and development, and water resource management. While the purposes of the water compacts vary, many of them deal with water allocation. To form a compact, two or more states agree. Each state legislature agrees. Sometimes congressional approval needed, if federal interests or balance of power is an issue. Of 59 compacts reviewed by the U.S. Government Accountability Office (GAO, 2007), 46 relied on commissions and the others on state agencies for administration.

Perhaps the most famous U.S. interstate water compact is on the Colorado River, which is explained briefly in Chapter 10 along with several other commission-based organizations. The Colorado River Compact has a long and much-studied history, which is centered around the original 1922 compact negotiated by future U.S. President Herbert Hoover.

The famous Colorado River Compact was signed in 1922 under the leadership of future-President Herbert Hoover.

From a background of administering a court decree, I can provide details on how another, smaller interstate compact is structured. This Pecos River Compact covers a river that rises in north-central New Mexico and flows into West Texas to Río Grande River, which forms part of the U.S.–Mexico border. The basin has several large, federally-constructed reservoirs and extensive irrigation is practiced. Controversy about water allocation led to New Mexico's 1973 suit in the Supreme Court and resolution required several Special Masters and the eventual appointment in 1988 of the River Master, which is the function that I have administered. The compact is administered with an annual delivery obligation that is set with an "inflow-outflow method" to correlate required water delivery with hydrologic indicators in the basin. The method was established through extensive negotiation in the Supreme Court case, and can only be modified by the River Master in consultation with the states. Appeals are to the Supreme Court.

The case shows that allocation formulas at transboundary points require careful engineering, hydrologic and legal studies and negotiations. The conflict resolution requires effective institutional arrangements that may be based on a commission with provisions for tie-breaking votes. In arid areas, the input-output method can be useful to compute water production from tributary basins but other methods are also available. In irrigated areas, significant conflict can arise over relatively small amounts of water because the economic consequences of shortages are high.

At the international level, water allocation involves riparian states. One of the most famous examples is the Nile River Basin, which flows through numerous African countries and discharges to the sea after supplying the water lifeline of Egypt. Other countries involved include Ethiopia (source of the Blue Nile), Sudan (where the White Nile and Blue Nile converge), and upstream states that include Burundi, the Democratic Republic of the Congo, Kenya, Rwanda, Central African Republic, and Tanzania, and Uganda.

Historically several Nile River treaties have been signed. The current Nile Basin Initiative is a partnership organization through which the riparian countries are establishing a commission to coordinate the use of water from the Nile River basin. The Initiative is supported by development partners, which include: African Development Bank, Japan, Canada, Netherlands, Denmark, Norway, European Commission, Sweden, Finland, Switzerland. France, UK, Germany, United Nations Development Program, Global Environment Facility, United States, Italy, and the World Bank (Nile Basin Initiative, 2010).

Other famous examples of water allocation among riparian states include the Tigris-Euphrates, Ganges, Indus, and Jordan River (including West Bank aquifers).

Transboundary water quality and environmental issues

After water allocation, water quality and environmental issues across borders might be the next most contentious category of issues. Many examples of problems exist, and a large-scale example is the La Plata River, which involves Argentina, Bolivia, Brazil, Paraguay, and Uruguay. Large scale river basins like the La Plata involve numerous economic, social, and environmental issues among the member states and negotiation of a treaty comprises a shared effort to deal with them together.

In the U.S., the Ohio River Valley Water Sanitation Commission (ORSANCO) was established to deal with interstate water pollution control. It is explained in Chapter 10.

As a state water quality regulator I was involved in interstate water quality disputes and learned that reaching agreement about them is very difficult. A significant case involved a coastal river that drained part of Southeastern Virginia into Albemarle Sound, which is an estuary located in Northeastern North Carolina. It is a long story, which is detailed partially in (Grigg, 1989) and the current status of the estuary is explained in (Albemarle-Pamlico National Estuary Program, 2010). Lessons that were evident to me during attempts to resolve interstate disputes were that point pollution sources in an upstream state are under the control of other regulatory authorities and are difficult to influence by downstream states, that it is hard to prove the causes of pollution, particularly biological effects, and that it is better to try for cooperative and long term approaches than to devolve into adversarial scenarios.

Lakes that straddle borders

Around the world are found many lakes that touch multiple borders and must be shared. Examples include the Great Lakes between the U.S. and Canada, Lake Titicaca along Peru and Bolivia, the Aral Sea between Kazakhstan and Turkmenistan in South-Central Asia, and Lake Peipsi, a shared waterbody between Russia and Estonia. Management of the Great Lakes involves cooperation through an International Joint Commission (see Chapter 10) and a Great Lakes Commission, which is an interstate compact agency with Ontario and Québec as Associate Members. According to Rogers and Hall (2003), management of Lake Peipsi illustrates how political will and cooperation can work as well as the difficulty of involving local citizens in sensitive political negotiations and the consequences of neglecting transparency and participation.

A river that forms a boundary

Rivers make convenient boundaries between political units of government, but they sometimes introduce problems due to shifting banks, confusing rights to the water, and other issues. The issue is the delineation of land title boundaries at the interfaces of land and water in rivers, lakes, estuaries, and the ocean. Issues that arise include legal principles defining property boundary movement, such as high– and low–water marks and change in physical location due to changes in land or water. Technical issues such as scour and sedimentation, dewatering of land, and covering of land by water result in dynamic changes in boundary. All of these create substantive water management issues about navigation, discharge permits, and allocation of water rights (Flushman, 2002).

One case that illustrates issues that arise is the Chattahoochee River which for part of its course forms the boundary between the state of Alabama and Georgia. The river has been the focus of a long-standing "water war" that mainly occurs over water supply for the growing mega-city Atlanta. The authority to allocate water along the river's course where it forms the boundary is one of the elements of uncertainty that arises. In this case, the boundary is defined as lying along the Alabama shoreline, and that seems to give Georgia greater rights to the river. This issue has not created major problems and seems less daunting than the general allocation of river water.

Cross-border aquifers

In a sense, every aquifer creates transboundary water issues, but conflicts are not as immediate as they are in surface water because groundwater moves slower and is more difficult to measure and assess. UNESCO (2010) has published a map of transboundary aquifers and in 2008 held a symposium in Greece on Transboundary Surface Waters and Shared Aquifers. They estimated that some 80% of water resources in the Mediterranean region involve transboundary issues, and that in North Africa and the Middle East, transboundary groundwater is the most important source of freshwater.

In the U.S., the giant Ogallala aquifer is a high-profile example of a transboundary aquifer. The Ogallala Aquifer covers a six-state agricultural region that experienced rapid growth in irrigation from the 1930s. By 1980 the aquifer was in decline and Congress authorized a comprehensive study (High Plains Associates, 1982). The outcome of the study was six alternative management strategies, but no action has been possible due to the patchwork nature of U.S. water law. However, local water districts and farmers have made improvements in water management and major problems remain spotty.

Complex issues crossing multiple borders

Perhaps the ultimate transboundary issue is the crossing of several boundaries by a large river. Examples around the world include the Nile, Danube, Rhine, and Mekong rivers, among many others. In the U.S., cases include the Colorado, Delaware, and Missouri-Mississippi rivers. These cases produce multiple issues or water allocation, water quality, and basin development.

For example, the Danube River requires an integrated, basin-wide framework for protecting water quality with a large number of riparian states, which actually add to 19 countries (Wolf and Newton, 2009). The Danube River Protection Convention was established as the first Regional Programme funded by the Global Environment Facility and the International Commission for the Protection of the Danube River is presently carrying out river studies. The Danube case, along with several other large and complex river systems, is discussed in Chapter 10.

Interbasin transfer

Interbasin transfers of water are attractive for those needing water, but fiercely resisted by those losing the water. Water transfers can be large scale, as the South-to-North water transfer project in China or the California State Water Project, or they can be small scale, as a diversion of municipal water from a stream with the return of wastewater to a different creek. Cases of interbasin transfer are presented in (Schaack and Anderson, 2001).

In the U.S., interbasin transfers are generally permitted under water law, and they are handled under water law rules. In the Colorado River Basin, the Colorado-Big Thompson project is the largest interbasin transfer and has impacts all the way along the river system. In the East, numerous transfers exist, such as a water supply for Virginia Beach, Virginia, where the city pipes water from the Roanoke River and diverts it away from Albemarle Sound. In a U.S. groundwater case, water developers proposed to pump large quantities of groundwater from the San Luis Valley to the Front Range, but their proposal was struck down in court.

CONCLUSIONS

Transboundary water issues occur when water moves across political jurisdictions. These are common with scales that range from simple to extremely complex. They test the ability of water managers and governance agencies to integrate solutions for the benefit of all. In addition to normal water conflicts, many inter-jurisdictional conflicts also occur.

Many examples of transboundary conflicts can be used to illustrate the problems at all scales. A classification framework provides a way to assemble scenarios of them that include water allocation across borders, interbasin transfers, water quality issues, international lakes, and even cross-border aquifers. These scenarios involve many different technical problems, but they share the institutional dilemmas of transboundary conflicts where lack of sovereign authorities to resolve disputes creates gridlock.

Shared governance is needed to resolve transboundary conflicts. It requires policy for negotiations among multiple authorities; empowerment of the coordination mechanism to balance water–sharing; formal inter-jurisdictional agreements; and appropriate controls based on planning and coordination and with regulation for compliance with the agreements. Inter-governmental agreements are needed to coordinate the issues between parties, such as inter-local agreements, interstate compacts, and international treaties.

Transboundary conflicts illustrate the main dilemma of IWRM: Once you move past the scale of local shared interests in solving a problem, you multiply the potential for misunderstanding and conflict. Local shared interests benefit from clear views of problems to be solved and the hoped-for desire to get along with neighbors. Both of these attributes are lacking once the scale moves past the local level and the distance between stakeholders moves past the level of neighborly relationships.

Postscript

Through my work as a Supreme Court-appointed River Master, I've been closely involved with a long-standing transboundary water conflict. Perhaps the most vivid lesson from this and related experiences is how common these conflicts are and how difficult they are to resolve, due to the many players and different incentive systems.

The earliest interstate conflict I can recall is what some of us characterized in the 1970s as "North Carolina-Virginia water issues." At the time I was the Assistant Secretary for Natural Resources in North Carolina, and was in frequent touch with our counterparts in Virginia. I worked closely with John Morris, who recently retired after many years as Director of North Carolina's Division of Water Resources. Our work showed us that there were three win-win issues on which we could cooperate with Virginia: resolution of a paper mill wastewater issue, management of a depleted interstate aquifer, and control of an interbasin diversion of water to Norfolk-Virginia Beach urban area. I would summarize our results with this by saying that these issues were far more difficult than we ever imagined.

The story is told in more detail in (Grigg, 1995) but a short version follows. We thought that by linking these three water issues an interstate agreement could be reached to meet everyone's needs. Perhaps it was the need of the City of Virginia Beach for a supplemental water supply that drove a wedge into the effort. Virginia Beach has grown from a small beach town of 22,584 in 1940 to well over 400,000 today. Its water supply situation was precarious and it relied on Norfolk for raw water and on desalting and wells. Droughts made Virginia Beach aware of its water vulnerability and compelled it to seek a more adequate and independent water supply. It studied its options and determined that a Lake Gaston interbasin diversion from the Roanoke River Basin would serve it best. This set into motion a process of conflict that took around 20 years to resolve. Virginia Beach needed North Carolina's cooperation, and the states had organized a bi-state committee to discuss our common water problems. In 1983 high-profile Hunt Senate race was on, and North Carolina announced that it would oppose the project, and that ended the cooperation.

To make a long story short, Virginia Beach applied for a 404 permit in 1983. After a number of federal studies, lawsuits, and political actions, Virginia Beach was finally successful and began diverting water in 1998. This case illustrated how solving the interstate transboundary water issue was only possible in the end through the court system. At the same time, the water quality issue is controlled by the Clean Water Act, not negotiation and cooperation, and no comprehensive solution to the groundwater issue has been possible.

The North Carolina-Virginia experience was repeated for me in the "Alabama-Georgia water wars," which are largely driven by the water supply needs of Atlanta, a rapidly–growing mega-city. I became involved while serving as a consultant to the Alabama Office of Water Resources, which was engaged in policy studies to develop new management and regulatory programs. The rivers that Atlanta depends on for its water are either tributary to or run along the border with Alabama. To make the situation more complex, the main river through Atlanta, the Apalachicola-Chattahoochee-Flint, discharges into an ecologically-sensitive Florida bay.

This conflict is much more difficult than the North Carolina-Virginia water issue because more players and issues are involved. The three states signed an interstate compact, but it broke down because they could not agree on an allocation formula for the water. The three governors have been personally involved and a federal mediator has worked to resolve the issues but no success is evident yet. Many lawyers are engaged and it is almost like it is a social security program for them.

As we think about our ideals for Total Water Management (see Chapter 1), it is well to keep in mind the realities of conflicts such as those described above. It seems that Mark Twain was right: "Whiskey is for drinking, and water is for fighting."

REFERENCES

Albemarle-Pamlico National Estuary Program (2010). *Estuaries: Where Rivers Meet the Sea*. http://www.apnep.org/intro.html (accessed 23 June 2010).

Flushman B. S. (2002). *Water Boundaries: Demystifying Land Boundaries Adjacent to Tidal or Navigable Waters*. John Wiley & Sons, Inc., New York.

Ganoulis J., Duckstein L., Literathry P. and Bogardi I. (1996). *Transboundary Water Resources Management: Institutional and Engineering Approaches*. Springer, Berlin.

Grigg N. S. (1989). Modeling and monitoring for estuary management. In: *International Conference on Estuary Management*, June, Hamburg.

Grigg N. S. (1995). *Water Resources Management: Principles, Regulations, and Cases*. McGraw-Hill, New York.

High Plains Associates (July 1982). *Six-State High Plains Ogallala Aquifer Regional Resources Study: Summary* (July 1982). Boston, MA.

Muys J. C., Sherk G. W. and O'Leary M. C. (2007). Utton transboundary resources center model interstate water compact. *Nat. Resour. J.*, (47), 18–84.
Nile Basin Initiative (2010). *About NBI*. http://www.nilebasin.org/ (accessed 23 June 2010).
Oregon State University (2009). *Program in Water Conflict Management and Transformation (PWCMT)*. http://www.transboundarywaters.orst.edu/ (accessed 3 December 2009).
Rogers P. and Hall A. W. (2003). *Effective Water Governance*. TEC Background Paper No. 7, Global Water Partnership Committee, Stockholm.
Schaack J. and Anderson S. (2001). Transbasin water transfers. In: *Proceedings of the 2001 USCID Water Management Conference*, 27–30 June, U.S. Committee on Irrigation and Drainage, Denver, CO.
UNESCO (2010). *Transboundary Waters Management*. http://portal.unesco.org (accessed 24 June 2010).
U.S. Government Accountability Office (2007). *Interstate Compacts: An Overview of the Structure and Governance of Environment and Natural Resource Compacts*. GAO-07-519, Washington, DC.
Wolf A. T. and Newton J. T. (2009). *Case Study of Transboundary Dispute Resolution: The Environmental Program for the Danube River*. http://www.transboundarywaters.orst.edu (accessed 3 December 2009).
World Bank (2009). *Transboundary Waters*. http://web.worldbank.org (accessed 19 October 2009).

Chapter 14

Water law: the legal foundation for water governance

Of all the institutional arrangements explained in Chapter 3, the law itself is the focal point. This occurs because societies need a way to write down their agreed-on rules of behavior and relationships. In the field of water, this means that water law is the recorded set of rules for all aspects of its management.

While water law varies among cultures and societies, it has the common aspect of conforming to the needs for water management, including functions such as allocation, discharging wastewater, and controlling floods, among others. So, it is appropriate to study water law on an international comparative basis so that the best features of different systems can be adopted. This is the general approach being taken around the world, but water management and governance involve so many variations that we are a long way from a global model of water law.

WATER LAW: FOUNDATION FOR GOVERNANCE

If the centerpiece of institutional arrangements is a comprehensive body of law, then the foundation for water governance is water law. Figure 3.4 showed how at the highest level the Rule of Law and government authority are needed for successful water governance. Then, at the next level, you have laws that include the law of contracts and agreements, enabling legislation for different water management programs, and regulatory law and rules. In addition to the legal structure and formal statutory law, common law and case law are important in water governance. In the U.S., many judicial decisions also shape the practice of water management and governance.

Law is the foundation of water governance.

Fisher (2009) explained the connections between water law and governance this way: "Governance is a function that goes beyond management to include all of the arrangements—institutional, constitutional and legal—that contribute to how water resources are used and developed from all perspectives." He shows how

 Governance and Management for Sustainable Water Systems. By Neil S Grigg. ISBN: 9781843393467.
Published by IWA Publishing, London, UK.

law involves a combination of formal and informal arrangements and specifies responsibilities as well as rights. While Fisher links governance closely to law, he also explained that as a relatively new term it has come to be a multi-dimensional concept and a multi-level activity. He sees it as including both governmental and nongovernment functions and to focus on rights and duties. This explanation is consistent with the presentation of water management in Chapter 2 and governance processes in Chapter 3.

This chapter draws from the explanations of water management in Chapter 2 and water governance in Chapter 3 to describe the range of legal mechanisms for water governance. It explains legal approaches for water allocation, health protection, environmental law, land use law, liability law, corporate law, the organization of government agencies, and other categories of law. It also serves as an introduction to more detailed references on types of water law and how they work.

While water laws necessarily deal with pieces of the water governance puzzle, they can be used conjunctively to create an integrated approach to water management. Figure 14.1 illustrates this integrated approach by showing the major categories of water law on the previous display of water management and regulation functions (Figure 2.5).

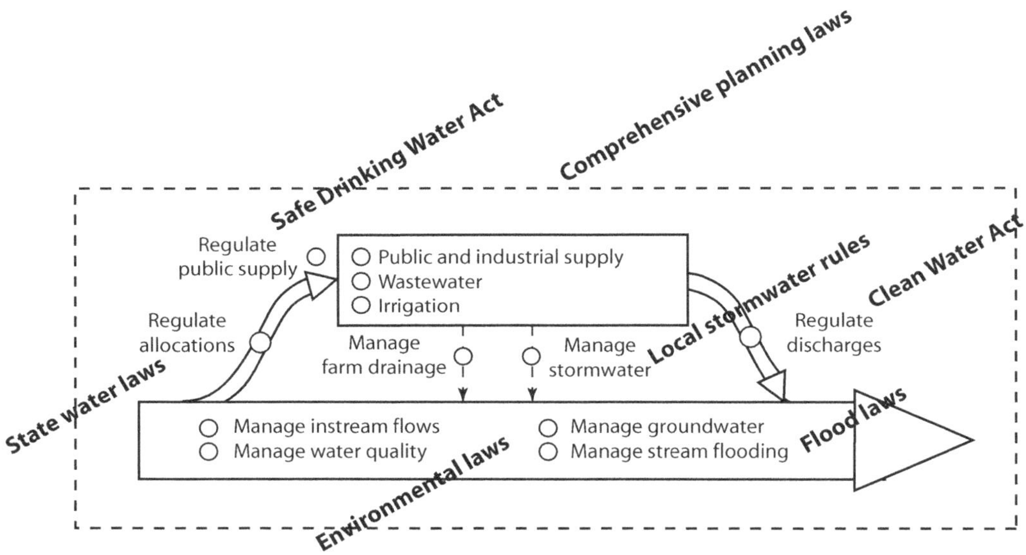

Figure 14.1 Water law aligned with management and regulation functions

CORRELATING WATER LAW AND MANAGEMENT

Figure 14.1 showed six major categories of water-related laws. Although the names of the laws are from the U.S. system, the categories of water law also apply in other legal systems. You can see these laws in action from Figure 14.2, which illustrates how they work along the flow path of a stream system.

The diagram shows specific points in the stream system where laws and regulations exert control. Comprehensive planning laws are meant to exercise integrated control over water management. Then, starting from the left of the diagram, you see that reservoir authorizations and rules on interbasin transfer (IBT) take effect. These special categories of law are explained later. Reservoir releases will be subject to laws such as for instream flows (ISF) and the Federal Power Act (FPA), and state water allocation laws will govern diversions from the stream and balance water uses among the competing demands.

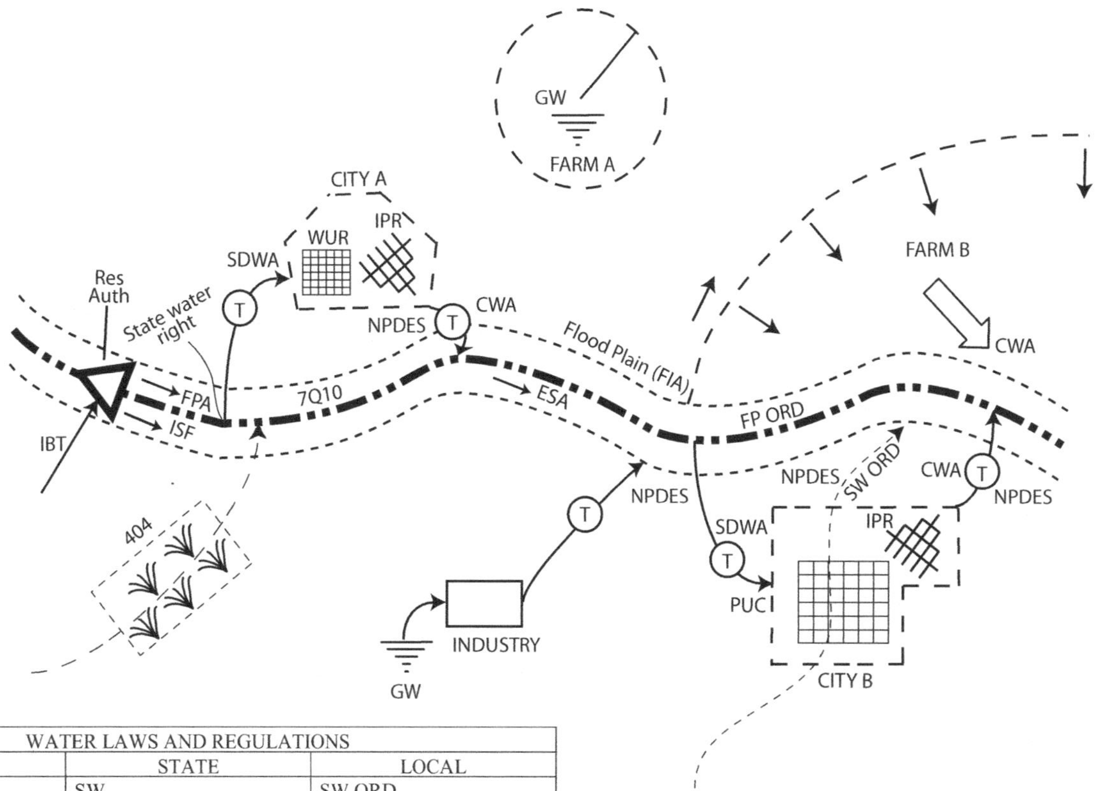

WATER LAWS AND REGULATIONS		
FEDERAL	STATE	LOCAL
CWA: NPDES	SW	SW ORD
CWA: 404	GW	GW
SDWA	7Q10	FP ORD
FPA	ISF	WUR
ESA	IBT	IPR
FIA	PUC	
NEPA		
Res Auth		

Figure 14.2 Law and regulations in surface and groundwater systems

Once water is diverted into a public water supply system, the Safe Drinking Water Act (SDWA) will govern water treatment, and local water use restrictions will govern operation of the distribution system. City B has a private water company that is also regulated for rates by the state's Public Utilities Commission (PUC).

Industrial pretreatment rules govern discharges to the collection system, as well as local ordinances. The wastewater treatment plant (WWTP) complies with the National Pollutant Discharge Elimination System (NPDES) permit program of the Clean Water Act (CWA). Section 404 regulations of the CWA will affect wetlands in a tributary area.

Also, seven-day, ten-year low flows (7Q10) and Endangered Species Act (ESA) rules govern streamflow for water quality and environmental purposes. The local flood plain ordinance will govern land use in the flood plain.

An industry and farm are shown pumping from wells and must comply with state or local ground water restrictions. The farm and the city must also comply with CWA rules on irrigation and stormwater return flows.

The acronyms of these water laws make an alphabet soup, but their occurrence is a reality of the highly-regulated and government-dominated world of water management.

Using the water management services and controls from Chapter 2, this table (Table 14.1) shows a list of the categories of laws that are shown on Figure 14.2.

Table 14.1

Public and industrial supply	See Cities A and B on diagram. Regulation of utility operations is by diverse controls, including the PUCs.
Regulation of water allocation	Water allocation law applies to all withdrawals from streams or aquifers. It also governs interbasin transfers. Regulation of releases may be by reservoir authorizations, which are by separate water laws.
Regulation of public supply	The main U.S. law is the SDWA. Local water use restrictions may also apply.
Wastewater management and water quality	See cities A and B and their wastewater systems. Regulation of discharges is by the CWA. IPR may also be used to control internal discharges to wastewater systems. Environmental water is regulated for quality by the CWA.
Irrigation and drainage	See Farms A and B on the diagram. No single law is applied to irrigation and drainage, and diverse controls affect these operations. Farm drainage is partially regulated by the CWA.
Stormwater	See City B and its local stormwater ordinance. Stormwater quality is partially regulated by the CWA.
Instream flow control	See ISF on the diagram. Some states have ISF laws. Environmental laws may also govern them. Water quality low flow rules also apply.
Groundwater management	See Farm A and the industry. Groundwater withdrawals are regulated by state laws.
Stream flooding	See the floodplain ordinance above City B. The Flood Insurance Act is the main U.S. laws.
Areawide and multi-purpose management	These are regulated by combinations of the other federal, state and local laws.

These categories of laws will differ among countries and even among states in the U.S. The next section provides brief explanations of how each category of them apply to water governance.

WATER LAW AND GOVERNANCE

Surface water allocation

Surface water allocation law is used to balance the withdrawal of water from natural streams. It creates rights to use water, usually under either a permit or appropriation system (Caponera, 1992). Under an ideal system, all water in streams would be accounted for and the diversions and returns would fall into an accounting system that served the needs of all users, including the environment. Of course, actual systems vary markedly from this ideal. Some work relatively well and others are practically non-existent.

In most countries, water allocation law is embodied in the national code, which will specify a system to allocate and control the use of water. In the U.S., water allocation is handled by state laws, and the states follow different approaches that generally conform either to the riparian doctrine, which is based on common law, or the appropriation doctrine, which assigns a property right to the use of water.

The more humid eastern states use the riparian doctrine, but as a practical matter, it gives way to a reasonable use approach, which normally uses permits or licenses as mechanisms to allocate the rights to withdraw water. So, while the law books may discuss the riparian doctrine, the practical situation is the use of permits and what might be called instead the "reasonable use administrative doctrine." The permit system can adapt to critical periods where the administrators determine reasonable alternatives to address low flow problems and to make plans for response.

The appropriation doctrine is used throughout the western states, which have mostly dry climates. It is designed to deal with the situation where there usually is not enough water to satisfy all users, so a system of allocation called "first in time, first in right" is applied. This doctrine is the main principle for water allocation in nineteen western states, although ten of these states follow hybrid approaches. In the appropriation doctrine, the water still belongs to the public, but the doctrine provides for the right to use it, in order of priority, as long as it is being applied to a beneficial use.

A water right must be initiated and perfected by appropriation and adjudication. A valid appropriation includes the intent to apply the water to a beneficial use, an actual diversion, and demonstration of application to the beneficial use. Water rights are administered by a system of rules and regulations, including calls on the river. Water rights can be lost through forfeiture or abandonment. Systems are necessary to transfer water rights to new owners or new uses. Rice and White (1987) describe how the administrative systems work in the State of Colorado, but the systems differ among states.

Many practical problems arise in administering the appropriation doctrine because of fluctuating hydrological systems and the general difficulty of accounting for water accurately. Imagine trying to determine each water right owner's entitlement in a stream that rises and falls according to hydrologic variation, with uncertain routing of flows from one point to another, unknown return flows, variable weather, and everyone diverting and releasing water according to schedules not under control of the administrators.

Governance issues with the appropriation doctrine focus on the concept that people can own the perpetual right to use water, regardless of what they use it for. A farmer might use practically all of a stream, for example, while environmental flows go begging and new arrivals find great difficulty in finding even minimum water supplies. The answer to these dilemmas, say water right owners, is to buy water rights from those who own them. Another issue is the management of water during periods of shortage. The doctrine of "first in time, first in right" means theoretically that many users will dry up while the senior right holders may get their full entitlements.

States are adapting to the shortcomings of the appropriation doctrine, and although it has flaws, it seems destined to continue in use in the West, with periodic tune-ups to respond to pressing needs.

Although water allocation law is governed by states, federal law can sometimes have the effect of overriding state law, as in the case of environmental requirements that must be met. In an extreme case, this might mean, for example, that to comply with a superior federal law, such as protection of a species of fish, a water right owner in a state could not use his entitlement. Problems like this are tested in court cases continually. Federal agencies must obtain state water rights or permits in most cases where they use water, and there is a category of federal reserved water rights for certain lands owned by the government.

The right to store water can be covered by permits as well. For example, in Colorado you can have a storage right for a certain volume of water, and it will authorize the diversion and storage of a certain amount of water each year. You would then be free to release and use your water according to your own schedule later.

Development of surface water allocation systems is proceeding in many countries. In China for example, the national water law was first approved in 1988 and contains a chapter on the Allocation and Economical Use of Water Resources (China.org.cn., 2009). It assigns to the State Council the responsibility for allocation of water on a nationwide basis and responsibilities for lower-level allocation to other levels of government. This approach links water allocation to central planning.

As another example, in Colombia they use an administrative approach and assign the role of water allocation to the regional autonomous corporations, who are to collect water use fees to improve waterways and manage the system (see later section with case study of Colombian water law).

The use of the permit approach opens the way for the administrative operation of water allocation systems that can accommodate contingencies, such as drought occurrence. Also, fees can be collected and used to improve waterways. For example, in Australia, where drought is a real problem, they recognize the need to reform the riparian system and move toward a water management system based on permits (or licenses). Objectives are to allocate water to the environment, allocate risk, create markets for rights, and make efficient use of water (Becker, 2008).

To develop effective surface water allocation laws studies are needed of stream and riparian systems, to include the needs of non-consumptive uses. The assignment of property rights to the use of water seems to increase conflict levels. People do not like administrative "czars" to control water use either, so no perfect solution is in sight. As always, cooperative approaches seem to be most promising but they require people to work together and to use shared governance systems, which are difficult to administer.

While a water czar might be efficient, people will doubt the equity of the arrangement.

Groundwater allocation and use

Although surface and groundwater systems are usually connected hydrologically, the law of groundwater has mostly developed independently of surface water law and is still evolving. Historically, groundwater law revolved around use of wells (Caponera, 1992). Early legal systems did not envision the need to control large scale aquifer systems, whether they crossed boundaries or not.

Groundwater law is a hybrid of water allocation law, resource law, and land use law. In the U.S., it is also based on state law and there is no national groundwater law other than the sole-source aquifer rule about protecting quality of drinking water sources. In Colorado, for example, different state laws regulate groundwater according to whether it is tributary to streams or not. The distinctions between different categories of groundwater are complex geologically and legally. In California, most groundwater use is unregulated but as in other states where this has been true, new rules seem inevitable.

To develop effective groundwater laws that work in local and regional areas, studies of groundwater hydrology and the stream-aquifer systems are needed. Conjunctive use of ground and surface waters can be an effective tool to integrate water management, but effective data and model systems are needed. Many improvements are needed in the governance of groundwater systems.

Drinking water safety

Safe drinking water law is focused on public health protection. International practice focuses on standards for drinking water quality promulgated through the World Health Organization (WHO) and through individual countries. Drinking water law generally provides for adoption of standards and for empowerment of the delivery systems to ensure that safe water reaches consumers.

In the U.S., the Safe Drinking Water Act (SDWA) applies health-based standards to protect drinking water against threats from improper disposal of chemicals, animal wastes, pesticides, human wastes, wastes injected underground, and natural substances. The law provides for a management system to implement a multiple-barrier approach to protecting public health.

Governance for safe drinking water law should focus on making it effective with management systems that work for different situations. In the U.S. and other higher-income countries, for example, waterborne disease outbreaks are rare, but drinking water systems still face many challenges (see Chapter 5). In many developing countries, the problem is not so much passing a correct law but in implementing and empowering management systems based on existing laws (see Chapter 12).

Environmental water quality law

The term "environmental water quality law" is used here to signal that it deals with quality of water in the environment, as opposed to drinking water quality. As explained in Chapter 6, the Clean Water Act (CWA) is the main US law governing environmental water quality, and it regulates discharges of pollutants into streams and also has related programs, such as protection of wetlands and regulation of dredging in streams.

Other developed countries have laws to control environmental water quality and, although they vary in their approaches, they have similar goals. In the E.U., the Water Framework Directive provides overall guidance (see Chapter 6), but many implementation issues remain to be solved.

As Chapter 6 explained, positive incentives and regulatory controls are required for a wastewater law to be effective. Features of the U.S. CWA address the major management issues that include regulation of point and nonpoint sources of pollution, monitoring of water quality, and enforcement of the law. These are complex tasks, however, and implementation of the law is the main challenge to governance. Effective law enforcement and continuing attention to appropriate levels of empowerment are required.

Environmental water laws

A nation's environmental laws can cover a broad swath of issues. The full range of environmental law extends beyond the scope of water management and reaches to concerns such as clean air and regulation of chemicals in the environment. The environmental water quality law addresses basic water pollution issues, but other laws are needed to address environmental planning and specific issues of habitat, instream flows, and related concerns.

In the U.S., the National Environmental Policy Act (NEPA), passed in 1970, establishes goals for environmental policy and the requirement for environmental impact statements (EIS) for major federal actions that affect the environment. A number of other laws add specific environmental provisions, which include:

- Endangered species protection: The Endangered Species Act (ESA) was passed in 1973 and amended several times. Arguably, it has had the greatest effect on water management of any of the environmental laws.
- Wild and scenic rivers protection: The Wild and Scenic Rivers Act was passed in 1968 to protect unique natural systems for preservation and outdoor recreation.

- Groundwater protection: the Comprehensive Environmental Response, Compensation, and Liability Act of 1980 (Superfund) provided for the cleanup of hazardous waste sites and led to many groundwater restoration projects.

In the U.S., instream flow laws are major tools to handle surface water issues. They are handled mostly at the level of state governments, and their purpose is to set aside water in streams for the protection of wildlife and public health. The law that requires minimum instream flows for water quality is based on that law, but there is an additional need for legal authority to coordinate among competing instream water purposes (see Chapter 8).

Normally, the water quality issue focuses on low flow conditions for the permitting of wastewater discharges. However, there may sometimes be environmental reasons to control flows during higher–flow periods, but these are normally not addressed by permits. The development of regulations to control instream flows is complex, and should involve low flow hydrology of study reaches, water quality and habitat impacts of low flow regimes, and estimates of required instream flows. These procedures should be followed by a search for common ground to establish minimum flows and the use of water quality, habitat, and future planning to determine the best approach to management of minimum flows.

Hydroelectric generation

In some cases, laws that govern other sectors are used to control water management. One of the best examples is law to control hydropower dams, which place a lot of stress on streams and water systems. In the U.S., the Federal Power Act (FPA), dating from 1920, provides for federal regulation and development of water power, and it controls licensing of nonfederal hydropower generation. One of the provisions of the act is the requirement for comprehensive planning of water resources as a condition for hydropower licensing.

Flood insurance and flood plain management

Laws to govern flood-related water management are closely aligned with land use (see Chapter 9). In the U.S., the flood management programs are controlled under the National Flood Insurance Act of 1968 and the companion Flood Disaster Protection Act of 1973.

Authorizations for water projects

Water Resources Development Acts (WRDAs) provide the basic authorization for federal projects, and other water management provisions added by Congress. These water laws specify the purposes of federally-sponsored water projects and contain provisions that lock in the management of stream systems.

Interbasin transfer law

Inter-basin transfers (IBT), or export of water from "basins–of–origin," are contentious because no basin wants to lose water to another. The original U.S. court case that legalized them involved a Colorado irrigation company and ever since then, interbasin transfer has been legal across the country. Being legal does not mean that people like it, and it generates conflicts almost inevitably whenever it is practiced.

Treaties and interstate compacts

Water flowing across state or national lines requires treaties, or interstate compacts, and lead to trans-boundary issues, which were discussed in Chapter 13. The law that governs these will be codified for special cases. That is, a state that enters into an interstate compact will approve it legislatively and Congress must approve of it as well.

For the most part, water laws are distinct within countries, but international water laws apply to flows across national boundaries and use of shared waters, such as the oceans. These are not as definitive as laws within sovereign nations. See Chapter 13 for more detail.

Public utility commission law

State law generally empowers public service commissions to regulate costs of water service for privately-owned utilities. These regulatory commissions stem back to about 1900 and the need to regulate railroad rates. Subsequently, this authority was extended to electric power utilities and a number of other services, including taxicab service. In most cases, water company rates are included in those regulated.

Water use restrictions

Local water use restrictions may be imposed during drought or other emergencies. This is a form of law where a local jurisdiction can decide on how to handle emergency allocation of its water supply or it can use its authority to change its rate structure to create incentives for water use efficiency. A number of court cases have challenged local decisions about water rates.

Stormwater ordinances

Stormwater programs are typically implemented by local governments and governed by their regulations. Chapter 7 explained how stormwater policy evolved through housing and subdivision laws and today has its own set of regulations and policies.

ROLES OF COURTS

It is surprising to note how much the third branch of government—the justice system—is involved in water resources management. The justice system involves federal, state, and local courts, as well as the administrative law system. Colorado has its own system of state district courts called "water courts," and the need for judicial understanding of water issues is widely-shared. Even justices of the U.S. Supreme Court are familiar with water cases, as I found out in a series of personal interviews as part of my work as a River Master.

CASE STUDY: WATER LAW IN COLOMBIA

A short case study of national water law in Colombia can add to our understanding of how water law and management should be aligned and how water law lays the foundation for water governance. The author worked with national officials and a team of advisors to study the national water laws of Colombia and how they can facilitate integrated water resources management. The following is a brief summary of the major findings (Grigg *et al.*, 2004). The brief discussion of Autonomous Regional Corporations in Chapter 10 also presents aspects of water law in Colombia. A description of the legal issues is also in (MacDonnell and Grigg, 2007).

Colombia has an extensive body of law related to water, which contains exemplary elements going back to the 1970s Natural Resources Code and its predecessors. However, the law is a patchwork and is located in many different legal instruments which could be harmonized to create a framework to implement integrated water resources management.

Colombia's Code of Renewable Natural Resources and Protection of the Environment was adopted in 1974 with rules for protection of the natural environment in a single code. It established guiding principles for renewable resources management with a stated objective being conservation, improvement and rational utilization of resources with equity, harmonious development, permanent availability, and maximum social participation. These desirable attributes of water law have thus been established in law for nearly 40 years.

The team's charge was to propose a framework for integrated water management and distinguish between elements of law and regulation processes. The strategy was to propose a framework for application of IWRM to the water–related goals of the country: to improve access to water; to improve water quality; to improve equity and public health; to improve effectiveness of planning and management systems; to improve water governance and inter-institutional coordination; to create a viable financing system; to control and protect environmental waters, including coastal and marine waters and ground water; and to introduce new mechanisms for urban drainage, flood control, and disaster prevention.

This project focused on the link between management and law and on institutional issues for integrating water management into the law. It did not examine basic legal doctrines, which were already established, but focused on urgent issues that included access to and reliability of water supply, safety of water supply, quantity and public health aspects of surface waters, environmental degradation, including in coastal areas, and better functioning of water–related institutions.

The two main areas of water law that we examined are water use law and water quality law. These can be applied to deal with special issues, such as preserved lands, wetland areas, and coastal zones.

The existing Colombian system of water law is based on user charges and the polluter-pays principle. The use of water requires a concession from the government and permission is required to discharge contaminants that harm human health or the environment. Thus, features similar to U.S. water law are included here except that in the U.S. these matters are generally handled at the state government level and in Colombia they are handled by Autonomous Regional Corporations.

For water use, the computation of charges and the method of collections are working partially. While Colombia is committed to a user-pay system, it has not been able to solve problems of water use allocation due to institutional complexities. Once again, it is not so much the design of the law as it is its implementation. Challenges include collection of charges, procedures for issuing concessions, conflicts among water users in periods of water shortage, and illegal use of water.

For water quality, there is lack of consensus about elements of the existing law. An obligation to remove 80% of certain contaminants seemed inflexible and inequitable, especially as applied to treatment of municipal wastes. A primary problem is the requirement for treatment of municipal wastes without the availability of the funding necessary to construct and operate required treatment facilities. Pollution charges for cities are regarded as taking away limited funding required to pay for treatment of wastes. Assessment and collection of wastewater charges is uneven and not universal.

Permission is required to take materials from riverbeds or beaches or to build structures in these places, similar to the U.S. CWA. An analysis of effects associated with proposed uses must be provided and conditions for environmental protection are included in the concession or permit, thus providing for environmental impact studies.

Elements that appear lacking in the law include management of floods and stormwater discharges, management of contamination from nonpoint sources, protection and use of coastal and marine waters, protection of ground water, and attention to environmental uses of water. There is no system for identifying desired instream uses of water or for protecting ecological values. There does not appear to be any provision for oversight of dam safety.

The team evaluated four proposals for a new water law in Colombia that follow existing laws. One of them offered a complete draft for a revised law, and the common element shared by the proposals is establishment of a special coordinating mechanism at the national level.

In any event, the team concluded that the new water law should: implement IWRM; align water management with national goals such as decentralization, democracy, sustainable development, and equity; coordinate and align existing laws and decrees; establish new principles and norms to guide water management in solution of urgent national problems; focus on the core themes of water quantity and water quality; include risk management topics, especially urban drainage and flood risk; include essential management support areas such as information systems, investigations, coordination, finance, and enforcement; and include interfaces with related areas, such as public services, coastal and marine issues, and environmental quality.

As always, the primary problem in the existing or proposed legal frameworks is mechanisms to implement the law across the sectors and regions of the country. The existing requirements call for plans at the national level, in the autonomous regional corporations, in regional governments (departments), at the water basin level, and at the municipal level. With a few exceptions, the needed coordination appears to be lacking on a widespread basis. Failure of this coordination is one of the main issues in current law.

CONCLUSIONS

As the recorded set of rules for all aspects of water management, water law provides the basic guidance for governance. It is much broader than a set of laws to deal with a specific situation, such as ownership of the right to use water, but each of its elements involves substantive issues of equity and responsibility. These elements provide the venues for stakeholders to work out functional systems to share and care for the resource.

Principles for water laws are known and disseminated widely across the nations, but the difficulty comes in their application. It is much more common to find a case where the written law and the actual practice differ than to find one where the law provides a guide for action that is followed closely by everyone. For this reason, water law must be accompanied by realistic and workable regulations and by a shared understanding among all stakeholders of how it should work.

Postscript

My introduction to water law was a graduate course at Colorado State University in 1967. It was taught by a local water lawyer, who explained the nuances of Colorado water law to us, with an emphasis on the appropriation doctrine. Since that time I have been fascinated by the overall subject and have had many experiences related to it. As an engineer I am a good target for the punch line of a joke about water law. "In Colorado, water law requires cooperation between lawyers and engineers. Some of the lawyers have become pretty good engineers, and some of the engineers have become pretty poor lawyers!"

It's true. Engineers need more training to understand the practice of law, but we can function as investigators, legal scholars, expert witnesses, enforcement officials and administrative judges, and I've worn all of those hats. Through these experiences I've gained a lot of respect for the legal process. Two experiences stand out vividly for me: serving as a River Master for the U.S. Supreme Court and leading a study of the water law in Colombia.

My 1988 appointment as a River Master came in 1988 when I was called by Supreme Court Justice Byron White, who informed me of my duties. I had been interviewed by the Special Master on the case, Charles Meyers, a distinguished attorney himself, and by the state officials of Texas and New Mexico, who represented the Attorney General and State Engineer offices of those states. I had to take a crash course in the case and prepare a River Master's report within a couple of months, which was a demanding experience, especially to gain knowledge of the legal nuances under which I had to operate. The case is discussed in a little detail in Chapter 13.

My experience in Colombia started in 2004 when I was contacted to work with the Department of Environment, Housing and Territorial Affairs and the World Bank to study the nation's water laws. Colombia was in the grips of armed conflict but the number of incidents was falling and it seemed like a good time for such a study. After all, water law can be one of the mediating institutions in society and might help in socializing some of the conflicting situations.

It was impressive to see how Colombia had already developed a *Code of Natural Resources* during the 1970s, and implemented water laws that were ahead of those in other Latin American countries. With this code and later laws and decrees, Colombia introduced systems for water ownership, water use concessions (or permits), and discharge concessions (discharge permits). It also introduced systems of charges for water use and for discharges, following the principle of "polluter pays."

During the 1950s, Colombia had organized its first "regional autonomous corporation," which was essentially a development organization for a particular region. The first one, for the Valle del Cauca region near the City of Cali, was modeled after the Tennessee Valley Authority. Colombia now has over 30 of these corporations, and has designated them as the country's main environmental authorities. It also has a National Environmental System (SINA), which emphasizes democratization, decentralization and sustainable development. SINA's participants are government environmental authorities, territorial authorities, departments (like state governments), municipalities, other environmental non-governmental organizations and private firms and farms.

As in many countries, Colombia's service delivery and public health improved dramatically in the last fifty years. Prior to 1875, the nation was emerging from the Colonial period, and services were basic, if they existed at all. In 1886, the first water supply system was inaugurated in Bogotá. It was only in 1957 that sewerage service began to be considered important. Water supply coverage in urban areas increased from 11% in 1938 to 29% in 1951 and to about 90% by 1990. Today, urban coverage is even higher, but rural coverage still lags.

As part of its regulatory strategy, Colombia has experimented with an innovative way to transfer income from wealthier to poorer utility customers. This is done through a system of classification of service zones by income. Higher income customers pay more and lower income customers pay less. Regulatory agencies regulate rates according to the proof by the public service authority that they are implementing this system.

Colombia has reached a crossroads in its use of water resources and in delivery of public services. As it recovers from its legacy of conflict and violence, it has many lessons to offer to its neighboring countries in Latin America, to other developing countries, and to North Americans.

REFERENCES

Becker J. C. (July 2008). *Australian Water Law Reform: A Case Study for Reform of Water Law Systems*. Kirby Lecture, www.aers.psu.edu/seminars/SemMats/Becker20081208.pdf (accessed 18 December 2009).

Caponera D. A. (1992). *Principles of Water Law and Administration: National and International*. Taylor and Francis (2007 edition revised by Marcella Nanni), New York.

China.org.cn (2009). *Water Law of the People's Republic of China*. http://www.china.org.cn/english/government/207454.htm (accessed 30 August 2010).

Fisher D. (2009). *The Law and Governance of Water Resources*. IWA Publications, London.

Grigg N. S., MacDonnell L. J., Fontane D. G., Howe C. W., Livingston M. L., Roesner L. A. and Salas J. D. (2004). *Integrated Water Management and Law in Colombia: Institutional Aspects of Water Management, Allocation of Uses, Control of Contamination, and Urban Drainage*. Colorado State University, Fort Collins, Colorado.

MacDonnell L. J. and Grigg N. S. (2007). Establishing a water law framework: the colombia example. *Water Int.*, **32**(4), 662–675.

Rice L. and White M. D. (1987). *Engineering Aspects of Water Law*. Wiley Interscience, New York.

Chapter 15

Regulation of water management

Water law must be translated into action through systems of regulations, which provide the basis for control and are normally the most visible of the governance tasks. Regulation takes many forms because it is society's way of controlling undesirable effects of business and other actions. It is handled through the processes of administrative law, which is the law that governs the actions of agencies of government. Examples of regulation include highway speed limits, bank regulations, and licensing engineers to practice. Obviously the number of examples can be very large in a complex society.

However, regulation has a negative connotation. People do not like to be regulated. In its worst form, regulation instills fear into managers who worry that they will be out of compliance and suffer penalties. Ineffective regulation also has a downside because it might induce disrespect for management goals and environmental needs.

It would be good if water management could proceed under the light touch of voluntary coordination, shared governance and other cooperative approaches, but it cannot operate fully without some actual control mechanisms. The trick is to get it right, to have the right balance between regulation and encouragement.

The theme of the book that governance requires effective institutional mechanisms is especially applicable to the topic of regulation. Systems of administrative law vary among countries, and the U.S. system is explained here to discuss the organization and functioning of regulatory processes for the water sector.

TYPES OF WATER REGULATIONS

In the water sector, regulation generally means to control compliance with water management rules where public impacts are important. These include most categories of water use and can be classified as:

- Regulations for health and safety (such as to supply safe drinking water)
- Regulations for water allocation (such as to recognize legal water rights)
- Regulations for environmental water quality (such as to maintain clean streams)
- Regulations for fish and wildlife protection (such as to provide instream flows)

- Regulations for rates and charges (such as rates of a private water company)
- Regulations for service access and quality (such as adequate water pressure)

REGULATORY PROCESSES

Regulatory processes begin with the development of regulations, called rulemaking, which must be based on formal legal authority if the regulations are to have force. In the U.S., after federal regulations are developed they pass into the Code of Federal Regulations and take effect. State and local governments have their own systems.

Once regulations are in effect, compliance mechanisms are required, including monitoring and reporting. It is also important to have effective enforcement mechanisms to ensure compliance.

Regulation is generally handled by government agencies and their public administrators. As public administration has extensive legal requirements, a water manager has many rules to follow and a regulator must follow set procedures in regulatory actions. These are set forth in administrative law.

The history of administrative law in the US falls into discrete periods and recent decades have seen many new regulations passed. As regulations increase, the sphere of action of water managers becomes tighter. Legislatures continue to pass laws, and more laws mean more regulations.

Administrative law comprises the constitutional and statutory provisions, court decisions, and directives that define rights and duties before administrative agencies, regulate procedures, and provide for review of actions. It has important features such as due process, public and interest group participation, freedom of information, open meetings, and other arrangements for transparency. Administrative law also structures rulemaking, adjudication, enforcement choices, and other decision making (Rosenbloom, 2003).

DEVELOPING REGULATIONS

Developing the regulations is the process of rulemaking, which implements executive, statutory or judicial guidelines. As an example of rulemaking, EPA's development of the Lead and Copper Rule (LCR) for drinking water will be explained. As EPA's strategy to control lead in drinking water, the LCR deals with the issue that dangerous levels of lead can be leached from distribution and plumbing systems. An interim standard in drinking water was established in 1975 and the 1986 amendments to the SDWA banned lead in new solder and use of leaded pipes in public water supply systems and plumbing. The 1986 amendments also directed EPA to review lead and copper regulations. The 1991 LCR introduced a new approach to regulating lead in drinking water at the tap.

The 1991 LCR required water suppliers to optimize treatment to control corrosion in customer's plumbing. The monitoring strategy was to sample water at the tap, and an action level of 0.015 mg/L (90th percentile) triggers a set of required responses.

The rule making for the LCR was extensive (Implementation, 1991; Powell, 1999; Regulating Lead, 1989). A number of institutional issues were faced, including EPA's decision to protect customers but not to have utilities be responsible for plumbing systems they could not control.

The science debates were about health effects due to lead exposure. Policy analysts thought that estimating exposure was complex because data on water at the tap and on consumption patterns were poor. Technical analysis was important in setting the LCR standard. The rule makers had to deal with complexity in water chemistry, how aggressive water is in attacking lead, and uncertainty about exposure and health effects data. The action level approach was a staff compromise in the end. As a result of a problem in the District of Columbia, EPA reviewed the LCR later. In Washington D.C. more than two-thirds of 6,100 lines sampled in 2001 showed lead levels above the action limits (Washington's, 2004).

Although a careful approach was taken to implementation of the LCR, indications today are that the problems are not over. The authority to enforce the regulation is diffused, and even main issues, such as not selling leaded solder, may not be under control. Many miles of lead service line remain in place, and it is not clear how long it will take to resolve these many sticky issues.

MAPPING LAWS INTO REGULATIONS

The regulatory process as implemented through government agencies is based on the authority of law. Therefore, for each type of regulation there must be an authorizing statute and an implementing agency. In some cases, there may be judicial decisions that interpret and/or reinforce the statute.

Figure 14.1 showed the categories of water law and how they apply to the processes of water management. Table 15.1 takes the mapping to one further level to provide examples of how in the U.S. system regulations are used in water management processes.

Table 15.1 U.S. system regulations and the water management process

Management process	Law	Regulatory control
Water resources planning	NEPA	Environmental impact statements
Water storage	Reservoir Authorizations	Reservoir operating rules
Water allocation	State water laws	State engineer rules
Drinking water quality	SDWA	Disinfection by-products rule
Environmental protection	ESA	Designation of threatened species
Flood control	Flood Insurance Act	Administration of flood insurance program
Hydropower generation	Federal Power Act	Relicensing rules
Wastewater discharge control	CWA	NPDES permit regulations

Enforcement

Enforcement requires mechanisms to respond to violations of regulations. The theory of environmental enforcement stems from law enforcement, and procedures had to be developed after environmental legislation took effect in recent years. On the other hand, police enforcement has been well-established for a much longer period.

How to balance incentives and regulatory controls is not always clear, especially when the groups being regulated have a hard time complying. The small water system problem is an example. Do you use persuasion or punishment when the small systems lack the capacity to handle all regulations?

In any event, there is a regular process for enforcement, such as the one explained by USEPA (2010) for the SDWA. The responsible officials will normally follow an enforcement management system or policy that includes:

- Notification and/or discovery of violation
- Investigation
- Initiation of statutory procedures (such as under the SDWA or CWA)
- First response (generally informal action such as technical assistance or letter)
- Formal enforcement action when needed

If violations pose an immediate risk to public health, the states or USEPA might proceed directly to a formal action. The types of formal enforcement actions are either on the civil side with administrative orders (with or without penalty) or judicial actions, which can include civil or criminal court cases.

REGULATING REGULATORS

Regulators must be under control as well as they systems they regulate, but how do you regulate regulators? This important question extends to all kinds of regulation, medical practice, financial institutions, and environmental control, among many others in our regulated society.

There is no perfect answer, but balance is needed. If regulators are too lax, systems can get out of control and harm the public interest. If they are too zealous, they can harm innocent people and the management systems that meet our needs.

My own interest in this issue was kindled when I was responsible for a state environmental regulatory program. The regulators in the organization ranged from on-the-ground inspectors to higher level executives. By and large, they wanted to do good jobs but their work was made complex by several issues. They were nervous that they were under a layer of political control. Sometimes they lacked guidance for appropriate levels of regulation and enforcement. Just as in any job, they needed feedback about their performance.

It seems to be the same general issues as "who regulates the police?" A police department can become corrupt and out-of-control. It needs oversight and performance review. Police work involves many laws that are familiar to people, such as traffic rules or laws against violent behavior. In contrast, environmental rules may not always be understandable to people.

Environmental regulations impact narrow sectors and are like business rules in that the rule makers might come up with onerous rules and those regulated need relief mechanisms. These are usually sought through the political process. Environmental compliance monitoring and enforcement officials need room for discretion, and this might require a lot of judgment.

SELF-REGULATION AND INFORMAL REGULATION

At small scales, people might work out ways to share local water supplies, but at larger scales they cannot negotiate directly with each other. In both cases, systems to decide among competing uses are needed. If people can work out their own issues on the ground, this is always best and is an example of self-regulation or informal regulation. Elinor Ostrom's Nobel Prize in economics studied this type of problem (see Chapter 3).

The principle of subsidiarity embodied in the EU law and authorized by the 1992 Treaty of Maastrich is consistent with pushing regulation to the lowest possible level. There are many examples of where it can pay off in water management, such as when people work out their own issues in a watershed.

As the scale increases, the possibilities for self or informal regulation diminish because direct negotiations between affected parties are not possible. This problem was explained in Chapter 13 as it unfolds in transboundary issues.

CONCLUSION

Water regulation in the U.S. seems well-established, but some wonder if it can be sustained following the existing models. After its awakening in the 1960s, environmental regulation had its birth in the 1970s, some 40 years ago now. The general approach adopted, command and control, has achieved successes in some of the worst areas—such as cleaning up polluted sewage—but more difficult areas such as nonpoint runoff have not improved nearly as much.

In these years, many lessons have been learned. Some of these were reported by Rouse () from his U.K. experiences, and explained in Chapter 3. They include:

- Need for government structures and capacity to facilitate sound integrated policies
- Separation of policy, regulation, and delivery

- Independent regulators empowered to act transparently
- Effective monitoring and enforcement of regulations
- Transparent information and reporting to combat corruption
- Public participation for transparency, buy-in and commitment
- Empowered utility operators free from politics to provide efficient services.

These are excellent lessons for certain scenarios of regulation, when it can be independent, empowered, and effective. Environmental law specialists Schoenbrod et. al. (2010) argue for a new model that might be more successful with small scale problems, where myriad issues are not solved by the prescriptive command-and-control approaches that are in practice in the U.S. today. Their ideas focus on instilling personal responsibility to care for the environment with tools where government shares power with the public. The idea would be to draw in people and groups who care about an issue and use the incentives to stimulate better environmental performance from everyone. Consumers and investors can demand better environmental performance from business and industry, and citizens can demand more from all levels of government. Building and vehicle owners can do more to green up their use of resources. These are heady ideas but need to be demonstrated for specific water governance issues.

Postscript

As an independent person, I never liked to be regulated. You might relate to that feeling as many people are that way. Whether it was discipline at home or school or even having to obey traffic rules, regulation seemed to work against freedom. But that's a narrow view. Regulation is good for the individual and for society, but it must be administered with an even hand.

I had always been aware of government rules, but I started to see new aspects of regulation when I worked in the environmental agency of North Carolina's state government. Although our political leadership was moderate and environmentally-sensitive, it was very tuned in to the business community and wanted to lean over backwards to avoid over-regulating. Political pressure was always to cut red tape, help businesses and smooth out regulatory controls. These were good things, as long as they did not get out of hand.

I was able to see that this pro-business orientation was necessary because the regulators working as government employees were risk-averse and always wanted to add to the rules for safety's sake. No one had an incentive to cut red tape. It was the other way around. This meant that permit waiting times got longer, requests for information increased, and when permits were finally issued, they were more stringent. I learned that not only does the process tend this way, the people who work as regulators tend to like the work because if they don't, they move on to other jobs.

I gained respect for our regulators, who had tough jobs and faced resistance. The head of our permit and enforcement section was a former city manager who wanted to do the job right. He didn't like to see political patronage in the agency and appreciated support to do a good and fair job. He had keen insight into the regulatory process and was a valuable public servant.

There were constant calls for "regulatory reform." I experienced this in other places as well, as people could see that regulations pile on top of regulations and tend to strangle us in red tape. This is evident in federal regulations, which in the U.S. are codified in the constantly-expanding Code of Federal Regulations. This ratcheting increase in regulations seems to call for a Year of Jubilee, a biblical concept for periodically revoking all past obligations.

As long as people have incentives to exceed the norms of behavior and performance that society needs, we will require regulatory processes. These integral parts of water governance require participants that begin with basic inspectors and end with judicial officials. It is along chain of processes that requires constant attention and tune-up.

REFERENCES

Implementation of the Safe Drinking Water Act (1991). Hearings before the subcommittee on superfund, ocean, and water protection of the Senate Committee on Environment and Public Works. In: *102nd Congress*, 17 May. Y4.P 96/10: Shrg.102-347 Doc.

Powell M. R. (1999). *Science at EPA: Information in the Regulatory Process. Appendix A: The 1991 Lead/Copper Drinking Water Rule.* Resources for the Future, Washington, DC.

Briskin J. and McCraw J. B. (1989). Regulating lead: an update. *J. Am. Water Works Assoc.*, **81**(7), 24–26, 123–126.

Rosenbloom D. H. (2003). *Administrative Law for Public Managers*. Westview Press, Boulder, CO.

Rouse M. (2007). *Institutional Governance and Regulation of Water Services: The Essential Elements.* IWA Publishing, London.

Schoenbrod D., Stewart R. and Wyman K. (2010). *Breaking the Logjam: Environmental Protection That Will Work.* Yale University Press, New Haven, CT.

USEPA (2010). *Drinking Water Monitoring, Compliance, and Enforcement.* http://www.epa.gov/safewater/sdwa/pdfs/fs_30ann_monitoring_web.pdf (accessed 4 January 2010).

Washington's water agency trying to get the lead out (2004). *ENR*, **252**(12), 19.

Chapter 16

Planning, coordination and shared governance

The practice of water resources planning is an important tool of water governance because the coordination it provides adds an element of control to water development and management. Acceptance of this control by water management authorities is normally voluntary, but it may be facilitated by incentives that provide benefits that come with cooperation. It may also be accepted better than heavy-handed regulation.

The term water resources planning is very broad and can mean many different things. I use the term to refer to a set of processes that range from drawing the plans for a facility to coordinating activities across a vast river basin region. The unique component of water resources planning is its use to coordinate the uses of water along a stream or in an area.

With so much conflict in the water arena, planning can serve as a peacemaker and coordinator. This role has been recognized by UNESCO (2003) in its program for Promoting Water as a Catalyst for Peace through Capacity Building, Research and Technical Assistance. This program targets decision makers and diplomats, water professionals, civil society, educators and post graduate students to explain how to enhance water security by strengthening capacity of stakeholders in the management of shared water resources by anticipating, preventing and resolving conflicts.

For this reason, planning involves many soft activities, such as public participation and dialog, but it also includes many scientific and engineering activities such as use of computers and simulation models to process real data and find plausible solutions to complex problems.

THE PRACTICE OF WATER RESOURCES PLANNING

Water resources planning is a broad and inclusive concept that includes any kind of plan-making, coordination, consultation, and cooperation. It may or may not end up with plans to build facilities and in many cases it is simply an ongoing dialog among stakeholders about shared problems and intentions.

The structure that water resources planning brings for water development and management is illustrated by concepts such as master planning and operations planning. For example, a water utility might plan for expansion of source water facilities or to upgrade and improve its water distribution network. By presenting a long term plan, it signals its intentions to governance authorities and lays a foundation for funding requests or regulatory decisions. Use of plans to coordinate among entities illustrates transparency in planning, which is important to enable each authority to align its plans with those of others so as to harmonize the use of resources and facilities.

Governance does not require regulation of all activity and, in fact, that will not work. There must be cooperation and coordination to facilitate voluntary actions of water users and managers, as well as acceptance of regulatory controls. Voluntary actions occur through cooperation, coordination, planning, and shared governance, which are closely related and are important to the success of water management.

The chapter outlines the general principles of planning, coordination, and shared governance and draws from the other chapters to show their central roles in the overall process of governance.

WATER RESOURCES PLANNING

In the context of water resources management, planning has come to mean more than to create plans. Prior to about 1970, emphasis in the U.S. was on development of water resources projects and infrastructure. To support project development, water resources planning meant to anticipate needs, create alternative ways to meet the needs, evaluate them, and to carry out a planning process with the goal of developing facilities (U.S. National Water Commission study, 1972).

After about 1970 and the emergence of the environmental movement, water resources planning changed. The stage had been set during the 1960s, when new approaches to planning were embodied in the U.S. Water Resources Planning Act of 1965, which provided for planning at the framework, basin, and local levels. The processes and provisions of this Act had evolved from experiences of the 1930s and were well-thought out but did not anticipate political developments that terminated them in 1981. Experience showed that formalized planning processes can be expensive and ineffective and it makes more sense to organize planning around shared problems and interests than to have static plans at the levels of each basin or governmental unit.

The experience showed that if plans are mandated top-down and not based on a perceived need at the local or operational levels, they are likely to fail. On the other hand, if there is an urgent need of some kind, such as a polluted river or threatened fishery, stakeholders will come together and want to plan so as to get action started, and they need a process to structure their work.

The process of planning is important because it gets people talking to each other about their intentions and to attract them to participate in shared governance. So in discussing planning, it is well to explain the process of planning as well as the plans themselves. In the process of water resources planning stakeholders go through steps toward an agreed on planning goal. Unless the stakeholders have a goal, discussions will lack structure. Therefore, a focal point such as creation of a river basin plan will help in maintaining the process. The general approach will work in different contexts, such as a small watershed plan, a river basin plan, an estuary improvement plan, a regional facilities plan, or others.

The essential elements of this process of planning can be illustrated by the U.S. National Estuary Program. A large fraction of the world's population lives near coastlines and they place heavy burdens on water quality and ecology in estuaries and coastal waters, which harbor diverse fish and wildlife species but have fragile ecologies. Examples of U.S. estuaries include Chesapeake Bay and the California Bay-Delta region. The same management concepts apply to bays and inlets of coastal waters everywhere. UNESCO (1977) compiled a world register of rivers discharging into the oceans and showed that of 260 major rivers discharging into the oceans, only 25% were regularly monitored for water quality. The most polluted estuaries were in the industrialized countries where the most data was collected.

In the U.S., as elsewhere estuaries are threatened by development, and the US Environmental Protection Agency established a National Estuary Program under the Water Quality Act of 1987. Early studies showed that water problems in coastal areas fit a pattern that includes (Davies, 1985): nutrient enrichment, eutrophication and nuisance algae; threats to dissolved oxygen levels; shellfish bed closures; lost and altered wetlands; disappearance of submerged aquatic vegetation; threats from toxics; diseased fish and shifts of fish species; salinity intrusion; and groundwater problems.

USEPA (1988) prepared an Estuary Program Primer to summarize lessons learned in the Chesapeake Bay and Great Lakes programs. It recommends collaborative, problem-solving approaches with a process of problem identification, characterization, and phased management. The process includes the Governor's nomination of an estuary; the convening of a "management conference" by the USEPA Administrator; estuary characterization; a Comprehensive Conservation and Management Plan (CCMP); implementation plans; and continuous monitoring. These take place in four phases: planning initiative, or building a management framework; characterization and problem definition; creation of a CCMP; and implementation.

The Management Conference is an umbrella for action and a forum to seek the three "C's"—cooperation, compromise, and consensus. It targets elected and appointed policy-making officials from all governmental levels; environmental managers from federal, state and local agencies; local scientific and academic communities; and private citizens and representatives from public and user interest groups that can include businesses, industries, and community and environmental organizations.

Notice in this planning process you have stakeholders involved closely in the process, you have a staff who can maintain the momentum by doing studies and reporting, and you have a focal point goal, to create a plan of some kind.

Levels of water resources plans

Whereas the process of planning is important in itself, it is also important to record the results of planning, to make agreements and to push ahead with facilities and management arrangements when appropriate. These needs were anticipated in the ill-fated Water Resources Planning Act of 1965, which specified three levels of plans: framework plans, river basin plans, and implementation plans.

While this exact language did not survive intact, the concept of hierarchies of plans is useful and is found in many kinds of public and policy planning. For example, in a city government you have comprehensive planning, master planning, and project planning. In the case of water resources, you can have policy plans (framework level), basin plans (which are a keystone of shared governance), and facility plans.

Coordination and joint planning

Inherent in the issue of the Tragedy of the Commons (see Chapter 8) is the reality that no one manages the myriad of smaller actions of the water industry. No one is in complete charge of the shared problems and no czar can take control of all water issues, even if such an approach was attempted. The obvious solution to this dilemma is coordination and shared responsibility.

Whether the issue is regional water or wastewater services, protection of an estuary, a flooding problem, or the need to share scarce water in a basin, joint planning can be an important governance tool to bring harmony among the diverse interests involved. This means that coordination itself is the needed control mechanism.

The term coordination actually means to bring harmony among diverse interests and it this harmony is the quality needed to integrate water decisions and actions. It is like the "invisible hand" that the economist Adam Smith used to explain how economies are coordinated.

As in the estuary example given above, you need an authority to call meetings and initiate the process. This need is recognized in the E.U. Water Framework directive, which calls for a "competent authority" to administer the planning processes (Green and Fernández-Bilbao, 2006).

Success in joint planning requires that the stakeholders coalesce around some sort of shared interest or problem. Examples might include:

- Watershed improvement
- River basin
- Estuary restoration
- Lake recovery
- Aquifer overdraft
- Instream flow issue
- River corridor improvement

Now that the Internet is used so much to share information, it is natural that web pages can be used to facilitate shared governance by coordinating information.

SHARED GOVERNANCE

The concept of shared governance means to distribute power to decide on actions and is a logical next step after coordinated or joint planning. The term is used in different ways, such as the ongoing arrangements for shared corporate governance between labor and management or administration-faculty sharing of power in a university. In the context of water governance, the arrangements are more likely to be ad hoc and created to meet particular needs.

When the term is used to refer to power-sharing across political units it could mean an inter-governmental agreement between levels of government or an inter-local agreement among cities and water districts. In any case, it is necessary to have a formal agreement whenever possible because successive political leaders might not honor informal agreements of the past.

The term "shared vision" planning has been coined to express the concept that stakeholders come together at the beginning of a planning process to agree on goals and the current status of a system. It is, of course, an idealistic concept and often stakeholders will come to meetings without being willing to agree because they want to advance certain agendas.

CONCLUSIONS

As a tool for management and governance, water resources planning can add coordination and a degree of control to water decisions. The term planning means different things to different people, but in its broadest sense it means a range processes from drawing facility plans to coordinating activities across regions. It includes soft activities, such as public participation, and data-driven scientific and engineering activities. In many cases, the process ends up as just a dialog among stakeholders who work out their own decisions, hopefully aligned with those of others.

The practice of water resources planning has evolved into today's version, with its many soft activities. The emergence of the environmental movement gets credit for most changes, and they add much cost to activities which can be more efficient if decision making is centralized in a few players.

Experience shows that top-down plans do not work as well as when local stakeholders come together to solve problems. In that sense, planning is like regulation: if local people can work out their own issues then there's no need for high-level government attention.

The essential elements planning are well known. Normally, you will have stakeholders involved closely in the process, a staff to maintain momentum, and goal to focus the action. This process can apply to all levels of plans, from the lowest-level implementation plan to high-level framework studies.

The Holy Grail of planning focuses on the use of shared vision planning to facilitate shared governance in decision making. It is very helpful when these concepts are successful, and their use by regional multipurpose organizations or by ad hoc groups can foster integration over areas and purposes of water management (see Chapter 10).

Postscript

During the 1970s I became a fan of water resources planning because its tools and techniques just seemed to make sense. Planning promised to be a way to work out differences and make water resources management work better. Perhaps my attraction to it was rooted in an instinct toward peace-making rather than starting conflicts.

I had the good fortune to be invited by Victor Koelzer to co-teach the course *Water Resources Planning* at Colorado State. Vic had long experience with the Bureau of Reclamation and Harza Engineering Company, where he had been a chief planner for Venezuela's Guri Dam and the Chicago Deep Tunnel Project. His class was like a World Bank workshop on how to plan and do feasibility studies for bank-financed infrastructure projects. He was an experienced professional and outstanding engineer. I have his picture with Guri Dam on my office wall.

With many events changing the nature of water resources management, I changed the course focus to include a range of soft topics, such as environmental impact analysis, public participation and the coordination process. These incorporated the state government experiences I brought to the course, as well as the ongoing water conflicts in Colorado. The outline of the course is contained in my 1985 book on Water Resources Planning, published by McGraw-Hill.

At that time the course included my experiences in a master plan for water use for the Nile River and failed U.S. attempts to engage in river basin planning. It seemed that interesting cases popped up everywhere. They set the stage for my next book on water resources, which included a series of case studies.

Water resources planning continued to change. At the time, Colorado was in the grips of the Two Forks Reservoir issue (see Chapter 11), and in the same decade it saw the secret water buy out project where the City of Thornton bought a large number of farms to get their water rights. In the UK, the Thatcher government had initiated a water privatization program, and you could feel the water industry changing right under your feet. It was moving from a mostly-government, centrally-planned industry to become a more nimble and dynamic one. Water resources planning had to change with it.

Today, the course on water resources has changed more to embody a range of management issues, as well as planning. To explain it, I usually say that it is based on technical analysis, but draws in political science, economics, history, law and finance. While that is a mouthful, it shows the diverse topics that must be brought to bear to fathom water resources planning challenges.

REFERENCES

American Society of Civil Engineers (1984). *Social and Environmental Objectives in Water Resources Planning and Management.* ASCE, Water Resources Planning and Management Division, Committee on Social and Environmental Objectives, New York.

Davies T. (1985). *Management Principles for Estuaries*. U.S. Environmental Protection Agency, Washington, DC.

Green C. and Fernández-Bilbao A. (2006). Implementing the water framework directive: how to define a "competent authority." *J. Contemp. Water Res. Educ. UCOWR*, **135**, 65–73.

UNESCO (1977). *World Register of Rivers Discharging into the Oceans*. Paris.
UNESCO (2003). *Promoting Water as a Catalyst for Peace Through Capacity Building, Research and Technical Assistance: From Potential Conflict to Co-operation Potential*. Paris.
U.S. Environmental Protection Agency (1988). *Estuary Program Primer*. Washington, DC.
U.S. National Water Commission (1972). *Water Resources Planning*. Washington, DC.

Chapter 17

Water policy and politics

Water policy and politics are foreign to many engineers and scientists, but they are the levers that determine outcomes of water management and governance. They are the most powerful tools we have to move toward sustainable water management. You might think of policy as the collection of all laws, directives, plans and strategies and of politics as the set of tools to implement it.

Policy analysis is like planning. You identify the problem, set goals, identify alternatives for solution, analyze them, and recommend a solution approach. Instead of a physical facility, policy analysis will normally recommend a strategy and program of action.

Policy is a multi-level process that extends as high as the multi-national level and as low as the smallest organization. It should be churned continuously to adapt old ways of doing things to new challenges that it faces. For that reason, it is a vital part of water governance.

THE POLICY PROCESS FOR WATER GOVERNANCE

Setting policy is the highest level function of water governance. The term means a plan or course of action, such as one made by a government that is intended to shape its decisions. This should be understood to be at a high level. Thus, water policy is the general plan that shapes how the resource is managed in an overall way.

Politics is a related term, and it generally means the art or science of government or participation in political affairs. As it relates directly to water management, politics can mean either applying the positive force of government to water decisions or it can mean the divisive actions of inducing conflict by forcing narrow political agendas into the decision process.

The term water policy seems to promise that somewhere in the archives of legislation, rules, and plans can be found the guides to action to solve our water-related problems. This might be true if water governance could be effective solely on the basis of rules and government programs. The reality is, of course, that starting with good policy is only a first (but essential) step, and it is also essential to have a good implementation program.

Policy should be targeted at institutional obstacles to Total Water Management, such as (Viessman and Welty, 1985; and Rogers, 1993):

- Ineffective law (patchwork of law, failure of law to integrate surface and ground water, separation of water quantity and quality law)
- Regulatory gridlock and proliferation of regulations
- Fragmented authority and difficulty in coordination (diffusion of jurisdictions and split committees)
- Market failure (not recognizing that water is not a free commodity and difficulty in setting prices)
- Conflicting priorities (non-uniform evaluation criteria and difficulty in setting investment priorities and implementing projects)

Policy-setting is the highest-level governance process because it establishes the goals, objectives and strategies for water services. As an example, goal-setting is the policy process of setting targets, such as "safe water for everyone." Then, strategy is needed to determine how to translate the goals into achievable outcomes. However, it is much more complex than that, and at the national level it must start by goals for the individual sectors, each of which has its own policies. Only then can an integrated water policy be assembled. Then, policy must be developed for the different agencies and levels of water governance. While it is sometimes more difficult to implement policy than to make it, the policy process is the starting point to figuring out how to approach water governance challenges.

This short chapter introduces the policy process as it takes place at different levels and venues. The chapter could have been bundled with the previous one because planning and policy are linked, but the policy process usually involves different players and deserves attention on its own. The chapter also raises the topic of the role of politics in governance, and opens the door for more detailed discussions that might follow in future books.

NATIONAL POLICY AND LEGISLATION

Although water is mostly a local issue, policy must start with national legislation or executive orders because of the shared nature of the resource. Thus, the combination of all legislation, executive orders and judicial decisions comprises national water policy.

How policy is made will differ among countries. You can imagine situations that range from a tightly-centralized policy arrangement controlled by ruling elites to a chaotic situation in a country with a policy of decentralization and weak local governments. Where overall governance is weak, the water policy will also be weak and ineffective, of course.

Getting legislation passed is difficult and requires a lot of momentum and support. Many, many well-meaning pieces of water legislation fall by the wayside. As a member of several national policy committees for water and infrastructure, I have seen how these committees seek to influence water-related legislation that originates with congressional committees. In that role, they are part of water's iron triangle, which is illustrated by Figure 17.1.

The figure illustrates the roles of government water agencies, legislators and politicians, and water users and interest groups. These are in operation at all three levels of government, and they interact in many ways during the water policy process.

The national water policy committees will normally have diverse memberships that are drawn from different places and different interest groups. Then, when a piece of legislation is introduced that is of broad interest to the group, they can express support to political leaders. In that sense, the committees are operating in the role of interest groups but may draw their membership from government water agencies as well as the private sector. When participating in such committees, water managers should disclose their affiliations, of course, but this is a good way to exert public-private cooperation in the water policy process.

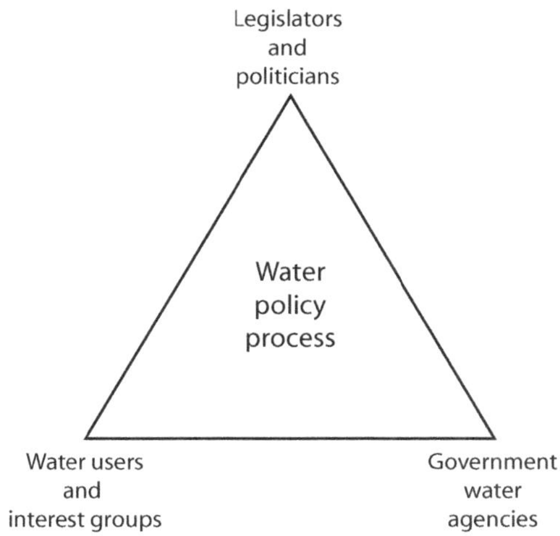

Figure 17.1 Iron triangle of water governance

Unless an interest group is in on the legislation from the beginning, it will be in a weak position to influence the outcome. Finding the balance in water policy is a difficult task because so many points of view must be represented.

STATE AND LOCAL WATER POLICY

State and local policies are the levels at which national policy is translated into action. Across the U.S., these vary sharply, depending on needs and the scale of the action. At the high end, states such as Texas and California have large presences in water policy, but smaller states where water is not a critical issue may only follow federal policies.

Local water policy deals with issues under the control of local governing bodies. For example, to implement the U.S. flood management program, policy is set by local political leaders and embodied in codes and directives of governing bodies. These set boundaries of flood plains, land use policies, financial plans, and related policies.

CORPORATE POLICY AND GOVERNANCE

Given the large numbers of water utilities, districts and other agencies, corporate policy and governance are important to sustainable water management. The roles of local governing boards are to set policies, make plans and strategies, and exercise corporate governance. It is important that these boards have competent and fair members, who work to learn about the complexities of water and who seek balanced decisions.

One of the interesting questions about local water board members is should they be elected or appointed? In the area where I live, both practices are in effect and, if members are elected, no one generally complains, but if they are appointed, it seems to attract complaints from one interest group or another. In Colorado, the board members of water conservancy districts (see Chapter 10) are appointed by state district judges, who may solicit applications, gather references, and hold interviews. One group that opposes this system is environmentalists, who might have a better chance of board selection through direct elections.

One argument for appointment is that the appointing authority can consider experience and balance in making appointments. On the other hand, in the election process the public is supposed to consider qualities like experience and judgment, but election campaigns are often won by the groups with the most money, energy, and incentive to gain control of a political organization.

Advisory boards are often used for governance and they offer a way to bring more experience and judgment to the decision process. I have participated in several of these and they typically hold regular meetings, review complex documents and plans, and make recommendations to governing boards. They offer a way to increase citizen participation in governance and also to attract greater expertise than elected boards usually have.

POLITICS IN WATER GOVERNANCE

Politics are an integral part of water governance because the operation of government and the relationships between people create the framework for work in the water industry. Politics is how government sets directions and policy guidance for decisions and actions. It includes levels of participation in government, including elections, interest group activities and passage of legislation.

When politics means applying the positive force of government to water decisions it helps lead to integrated decisions, but when it means the divisive action of pushing narrow political agendas, it works against good water management. That is not to say that there is no place for pushing extreme environmental values, but narrow agendas can also work against wise triple bottom line decisions.

Since water managers are inevitably involved in politics, the trick is to be involved in the legitimate processes of government but not in conflicts among politicians, inappropriate interest groups or political parties. The public servant must be involved in government but in the public interest. The basic level of participation is as an unbiased expert in policy analysis, which is appropriate and essential.

Governance should ensure that public involvement is a two-way process that occurs within the framework of democratic government. It should feature honest communication, based on mutual respect and a sense of interdependence. Activities include public information; task forces and advisory groups; public meetings, conferences and workshops; mediation sessions; collaborative problem-solving; and negotiation and arbitration.

Both elected and appointed officials work in politics. In some cases, they may drum up support for programs when the programs are not really needed. The goal could be to enhance images by enlarging programs and budgets. These larger programs result in higher employment, more powerful positions and higher pay. Sometimes problems are caused by agency staffers with strong views.

Even water quality standards or water conservation limits are political in that they involve the positions of engineers and interest groups. In storm sewer standards, for example, the city engineer may prefer larger systems to reduce maintenance and local flooding, but the developer may prefer cheaper ones. The result is often a tradeoff and politically–set standard.

Public administrators must understand political tradeoffs and explain options to elected officials. For example, budget politics are important to maintenance of water systems. In politics, there is a tendency to defer certain capital and maintenance needs, especially those that are not so visible such as pipeline repairs.

Water provides a lever to control land development, which is a source of wealth and influence. Land developers and industrial leaders support water management actions favorable to them. The water decision-making process attracts powerful people and business interests.

Solving political problems requires leadership highlighted by cooperation. Water problems require this more than they require the "push it through" style that was dominant in earlier days. Cooperation in water management is any form of working together to manage water, as in a region or river basin. Regionalization means integration or cooperation on a regional basis. Examples are explained in Chapter 10.

Using cooperative approaches reduces conflict and increases the likelihood of success. However, the political aspects of water decisions may result in conflict. Non-regulatory techniques to deal with conflict include negotiation, conflict resolution, and alternative dispute resolution.

Awareness of the need for win-win approaches in water management is increasing. The politics of the water planning process takes much more time and seems less efficient than non-participatory approaches. Thus, governance must prepare decision makers to spend the extra time and money required to reach wise decisions. Around the world leading edge thinkers are seeing that in addition to technical subjects, water education must address conflict management, dealing with critical events, decision making processes, working with the public and governmental bodies, policy analysis, and integrated approaches problem-solving approaches.

CIVICS AND WATER STEWARDSHIP

Total Water Management requires stewardship, which is a requirement of citizenship. Citizenship is fundamental to the success of water management and relies on positive action and people not always having to be regulated. Citizenship is a pillar of democracy and trust in government and everyone's responsibility. Governance should promote civics education to help people understand their rights for equity in public services and environmental resources; help them understand their personal responsibilities by explaining water conservation and water stewardship; help foster participation in democracy by organizing public involvement in planning and decision making; help people to share and interact in identifying solutions to water problems; and help people feel better about government by explaining environmental issues.

CASE STUDY: NATIONAL WATER RESOURCES POLICY

The Global Water Partnership has a case study on National Water Resources Policy. The concept presented is to set national goals and objectives for water management including regions, catchments, shared or transboundary water resources, and inter-basin transfers.

First off, it is recognized that subsidiarity is an important principle so that decisions are at the lowest appropriate level. Some of the decisions about policy are that it may include legal issues, political matters such as decentralization, economic incentives and criteria, capacity-building, and environmental monitoring and control. Some of the lessons learned are that policies are more useful if they are designed over the long term and not in the crisis mode, although it is recognized that a crisis may create opportunities for policy change. Also, a broad look is needed to go beyond the sectoral view and extend to macroeconomic and development needs.

Some key points made are that policy should clarify the roles of government and other stakeholders. The roles of government as regulator, organizer of the participatory process and adjudicator in cases of conflict require special attention. Policy should set priorities for water issues to ensure a focused policy, and make explicit links between land use and other economic activities. Policy should recognize that considering water as a social and economic good means allocating resources to the greatest value to society and start with basic needs. To have transparency, it is important that stakeholders are engaged in policy formulation and have mechanisms for conflict resolution.

CONCLUSIONS

Water policy is the strategy-setting part of governance. Politics is the art of using government mechanisms to implement it. Good policy is a necessary but not sufficient condition for effective water governance. It should be aimed at the institutional problems that block our success in water management, including problems such as ineffective law, regulatory gridlock, fragmented authority, and conflicting priorities.

Water policy starts at the national level and offers choices to conform to different cultural and economic settings. It is at lower levels where policy must be translated into action, so the implementing mechanisms must be functional.

Experience shows that interest groups call for "national" water policy and many papers and journals write about it, but policy comes in pieces as legislative bodies deal with issues one at a time. The process ranges from those in democratic countries which may lack coherent strategic plans to those in authoritarian countries where five-year plans may drive policy.

So, at the end of the day it may seem easy to make policy and it will be to write it, but the difficult part is making policy that works. This requires genuine engagement up and down the chain of responsibility so the policies not only make rules, they empower the water management system with the support and institutional mobilization it needs.

Postscript

It might be a stretch to say this, but to power junkies the term policy seems almost sexy. Perhaps another word to describe its attractiveness would be as a stimulant to get us excited about the use of power to solve vexing water problems. I felt that way about water policy in the 1970s, when I was able to begin participating on interest group national policy committees for water and infrastructure. These committees would meet and grapple with issues, then go off to congressional offices "on the Hill" to pass the word.

I soon became disillusioned with this process when I learned how little interest politicians had in most of our policy advocacy. It takes an enormous amount of such advocacy to change things, and they mainly wanted to know if your interest group was for or against an issue.

Things are different at the level of local or state politics. If you see a water issue and know how to solve it, you can access your representative and his or her staff and talk things over. Unfortunately, the water issues are usually so complex that for every gain you make, there is a corresponding loss. I've seen more than one newly-elected representative get excited about solving water problems with legislation, only to draw in their horns when they find out how complex it really is.

Setting policy for high-level issues such as those embodied in the Clean Water Act or in passing a rule, such as under the Safe Drinking Water Act, takes years and mobilizes many interest groups and power brokers. For every large scale initiative that succeeds, I would guess that hundreds fail. This has been illustrated by health care policy in the U.S., which has thwarted many an attempt at reform.

So, policy might be sexy, but the price to play in the game is high and your chances of winning are small. I can think of other analogies, such as winning the lottery, but at least policy depends on knowledge and analysis and not pure luck, with the odds stacked against you.

Perhaps the best lesson is to understand that in water management we play in a gigantic game with many levels. It is not possible to move policy at the upper levels very much or very quickly, but you can use appropriate policy instruments to solve water problems where they occur, at the local levels.

REFERENCES

Rogers P. (1993). *America's Water: Federal Roles and Responsibilities*. MIT Press, Cambridge.
Viessman Jr. W. and Welty C. (1985). *Water Management: Technology and Institutions*. Harper & Row, New York.

Chapter 18
Toward a water governance future

To end this discussion of water governance, we return to the realization that water problems cannot be solved in isolation but only through integrated approaches that operate within effective governance systems. Policymakers and managers urgently need these approaches to face the pressures and conflicts on water caused by population growth, economic and environmental pressures, and natural and manmade disasters.

In particular, water problems cannot be solved purely through private sector initiatives because too many public interests are involved. We can almost imagine that there is some kind of divine plot behind water conflicts that forces humans to work together on their shared problems.

These admonitions ring true, but they also seem abstract and complex and to need clarification through real-world examples and tests. Given this complexity, what is the future of water governance in a multi-faceted and rapidly-changing world,? Can effective systems be brought to bear to facilitate sustainable use of water, or is the future to be a continuation and even expansion of the state of confusion in water governance suggested by the World Water Assessment Programme (2007)? We can't know the future, of course, but we can think about it. That is the purpose of this chapter.

GLOBAL WATER NEEDS

While the world's water needs can be managed, a crisis will deepen unless effective management systems are used. Meeting human needs for water will always be the highest priority, but this does not mean that waste and despoliation of the environment can or should be tolerated. Now, it seems that no corner of the planet is untouched by human activity, and the urgency of protecting natural resources is much greater than in the past. Protection of watersheds, ecosystems, and other natural systems is urgent in the face on going threats from development. Balanced development and water use can lead to the sustainable strategies that are needed everywhere.

Developing countries face great challenges in meeting the needs for safe water and sanitation services for everyone, and this issue carries over to any place where people struggle to have access to safe and reliable supplies. Even in developed countries, effective water management systems may be available, but the countries still face challenges in balancing competing uses, promoting conservation, and empowering citizen and corporate stewardship of water.

The ultimate challenge in water governance is to balance uses of the resource. Every water issue can be framed in the context of competing demands, whether for withdrawal or instream use, for municipal water supply or irrigation, or other tradeoffs. The balancing of water uses requires that they be valued by their worth to society. This requires water governance to sort out the mixture of public and private purposes so that pricing and other decisions can be made. The market can work for allocation of some attributes of water resources, such as the right to use water for economic purposes, but other societal purposes cannot be solved by market allocation.

ELEMENTS OF GOVERNANCE

While the goal in the book was to make the discussion as uncluttered as possible, the concepts of governance and institutional arrangements are inherently complex. Figure 18.1 draws the discussion together into four straightforward question-and-answer sets to explain them and is named the "ABCs of water governance."

Water governance

What is it?	Why is it important?	How does it work?	Whose job is it?
It is the operation of society's institutions to manage water resources and services.	If it is ineffective, human and environmental water needs will be unmet.	It works through policy, empowerment and control of water institutions.	Government, social institutions, and individuals have roles in it.

Figure 18.1 The ABCs of water governance

The complexity arises through the word "institution," which appears three times on the diagram. Institutions are society's mechanisms to manage its affairs and comprise many types of arrangements. On the diagram the term is used in the following ways:

- Society's institutions to manage water resources and services—refers to all institutions, political, social, legal and economic
- Control of water institutions—refers to the subset of society's institutions that bear directly on water management
- Social institutions—refers to society's relationships, obligations, roles and functions

While these concepts do involve many interacting elements, they are the only way to explain how society goes about water management. It comprises much more than infrastructure and engineering management; it draws in the public, the law and the many control mechanisms that are required to make society function for people and the preservation of the planet.

The necessary conditions of successful water governance are good policy, capable organizations and workforces, adequate resources, and ways to blend regulatory strategies with volunteer actions to raise levels of compliance. Policy, control, and empowerment must be applied at different levels in different places. While the starting point for water governance is to set policies and pass laws, it is usually more difficult to implement them than to set them in the first place. The concept of subsidiarity is a good way to assert that water should be managed at the lowest government level where it works.

The degree to which these are met across countries and regions determines the effectiveness and sustainability of water management and services. The governance needed in a high-income country may look entirely different from that required in a low-income country that lacks basic institutions. However, certain attributes of water management and governance are valid across the spectrum of settings. These include government policy that recognizes needs for all water purposes and stakeholders and sets out mechanisms for control and empowerment to provide for them.

Governance should include appropriate control levers through regulation, planning, and coordination that respond the most urgent and important scenarios. Governance is broader than regulation alone, but regulation is a core process. It involves both voluntary actions of water users and managers and regulatory controls.

Equally as important are the empowerment mechanisms that respond to the needs of organizations and workforces to build the institutional, technical, and individual capacities needed for sustainable water management.

Models of governance are needed to promote shared, corporate, and self-governance to accompany government actions. Where cooperation succeeds, regulatory controls can be reduced and levels of public acceptance are likely to be higher. So subsidiarity, flexibility, local control and self-regulation are attractive concepts to promote effective water governance.

MATCHING GOVERNANCE ACTIONS TO WATER ISSUES

Each category of water management faces challenges to water governance. Table 18.1 gives a short list of the principal challenges:

Table 18.1 Principle challenges of water management to water governance

Public and industrial supply. The policy for access to safe and adequate water supplies for all is mostly in place but implementing it is difficult. Empowerment and control are needed in varying degrees across countries. Regulating the safety of water is critical and involves source, treatment, and distribution systems. With water scarcity, there is a need to balance supplies and work on water efficiency.
Regulation of water allocation. Balanced allocation of water taken from natural systems requires appropriate laws of water use and effective administration of water diversion systems. Allocation of sufficient water for low-income people and for environmental flows are continuing issues.
Wastewater and environmental water quality. Access to sanitation and wastewater service needs improvement in developing countries. Even many middle income countries need to implement cost-effective wastewater treatment systems. To regulate discharges into water bodies for health and environment requires diligence and commitment in all countries.
Irrigation and drainage. In developed countries, like the US and Australia, irrigation systems require a balance of supply with regulation of environmental impacts. In many developing countries, the issues of farm income and food security are more paramount. Systems for removal of drainage water from farms need to be developed and sustained and regulated. It is difficult for irrigators to pay full costs, but unless they do pricing mechanisms for water management cannot be implemented completely.
Stormwater systems. Stormwater and flood systems and nonpoint source controls are difficult to govern, even in high-income countries, and they become more urgent in developing countries where lack of control adds to misery, disease, and death.
Instream flow management. Access to instream flows for hydropower, navigation, recreation, fish and wildlife, and to convey water to users is needed in all countries, and managing them poses a stringent test to integrated water management.
Regulation of groundwater withdrawals and recharges. Problems with groundwater management are spotty, but failure to control use and quality can ruin aquifers permanently in critical locations such as islands, coastal areas, and industrial locations.
Areawide and multi-purpose management. Special purpose organizations can be effective for coordination, basin planning, and required water services, and their use should be considered in many water-challenged areas around the world.
Transboundary water problems. Transboundary water issues pose another critical challenge to water governance, and treaties need to be developed and enforced wherever serious issues occur.

Within these issues, the promotion of water efficiency is an important goal of governance. Whereas balancing is to move water uses toward high values and equitable allocation, efficiency is aimed at getting more value across the board and toward eliminating waste.

Disaster preparedness is also an important responsibility of governance. Water-related disasters, whether caused by flood, drought, or disease outbreaks, require preparation, response, and recovery.

The climate change discussion is like a wild card in the discussion of issues. It requires study of uncertainty and adaptation, rather than about deliberate changes in policy. Water policy is fragmented and the places to make changes are mostly at the state and local levels. The impact of climate change on water management requires local places to adapt to changes that they either already see or anticipate for the future. Adapting to the need for efficiency, rather than increasing supply, or to adapt to flood changes with better measures might be appropriate policies.

ROLES OF THE GOVERNANCE PLAYERS

The Global Water Partnership (2007) identified lessons about the roles of the players in IWRM and its governance. These provide a different view than the functional view given in the last section. The order of their discussion has been re-arranged to emphasize the importance of the local and river basin levels (Table 18.2):

Table 18.2 Roles of the players in IWRM and its governance

Service providers	• Structure is dependent on the culture and society • Works best with property rights and obligations • Require technical tools (pricing, audits, transparency, management systems)
Local government	• Require enabling environment • Local leadership is needed • Public information is required • Planning, action and recognition are required
RBOs	• Need trusted technical competencies • Focus on solutions to recurring problems • Requires broad stakeholder involvement • Require sustaining revenue • Require clear jurisdiction and powers
National apex bodies	• Establishment can be slow, requiring time to gain legitimacy • Effectiveness linked to historical context • All stakeholders must support bodies
Regulators	• Require adequate staff, capability and resources • Require effective statutes • Require sense of ownership of stakeholders • Require benchmarks to work toward • Require program of legal education and awareness building
Transboundary	• Once established, these and their agreements are robust • Need technical secretariats and to emphasize implementing mechanisms • Need respected broker to facilitate agreement • National policy should support coordination to facilitate success in transboundary issues
Civil society	• Should collaborate with service providers and local government • Must not be taken over by narrow interest groups

GOVERNANCE FOR THE PUBLIC INTEREST: THE PUBLIC-PRIVATE DEBATE

If any message about water management is clear, it is about its complexity. Mantras such as "water is a public good" or "privatization" ring hollow unless they are aligned correctly with the context and the problems at hand. Every issue is multi-faceted. For example:

- Water supply can provided as a commodity for sale, but low income people may not afford their essential basic supply
- Wastewater service can be provided by for-profit entities, but low income people may not afford the service and the private companies still must meet discharge requirements
- Irrigation water can be sold as a commodity, but large scale systems may require government assistance and small farmers need capacity-building assistance
- Flood insurance can be sold but requires a subsidy and it is impossible to charge people the full cost of flood protection.
- Instream flows cannot be unbundled as services for management by the private sector

FUTURE GOVERNANCE FOR WATER SERVICES

Seppälä (2004) examined the governance requirements for public services, with a focus on the situation in Kenya. His conclusions offer tips for needs across the spectrum of development needs:

- Reform processes should be locally-driven, not controlled by external support agencies and foreign experts
- Both public and private solutions may be appropriate, depending on the local situations
- Inadequate financial resources and related management structures continue to be a difficult and important issue
- Customer involvement needs more attention, as opposed to supply-driven policies
- Rural models of community participation may offer help for urban utilities
- Utilities should adopt future-oriented management changes

A FUTURE FOR INTEGRATED WATER RESOURCES MANAGEMENT?

The water crisis that is creeping onto us threatens our ability to meet a range of needs that begin with basic safe water and sanitation service for all. It reaches from there through the need for other social, economic and environmental water services, and includes a comprehensive environmental water quality control program. Within these, environmental water requirements demand attention to instream flows and a good balance is needed to provide effective flood disaster management and stormwater service, while respecting environmental impacts.

The challenge to meeting these needs in an integrated way is in our institutional mechanisms. Most problems present creeping challenges, like the persistent daily problem of providing safe water to all. However, some break out suddenly, such as the devastation of parts of New Orleans from the Hurricane Katrina flooding. Whether the problems creep up or break out suddenly, it is the job of water governance to anticipate and solve them, even if it is difficult to sustain a "crisis mentality."

Of the some 180 countries in the world, most lie in the middle- or low-income ranges. Their issues range from urgent problems of access to water and sanitation to the ongoing problems of managing and maintaining infrastructure. Institutional capacity-building is the biggest step needed in almost all cases and will determine success or failure in providing wastewater treatment, renewing irrigation systems in rural areas, protecting people from flooding in cities as well as rural areas, and gaining control of environmental issues.

Although in developed countries most water challenges seem less-daunting than some of today's other intractable issues, they will require continued discipline and commitment to provide water for environmental systems and instream flows, to regulate drinking water safety and wastewater disposal effectively, to gain control of nonpoint sources, to improve equity in allocation of water, and to integrate the management of water services. Even these countries cannot afford to be complacent or to think that all problems have been solved. The financial crisis has set renewal of aging infrastructure back, and institutional barriers to reforming water management are found everywhere.

The stages of water management can be portrayed as shown on Figure 18.2, which illustrates relative priorities of water services as they change along the axis of institutional development. The top service—safe water supply—changes to become more accessible, affordable and dependable, while other services come along later. The services of stormwater and instream flow management are vital, but they are normally put in place after basic services such as water supply and wastewater management. You would expect, for example, that a lower income country would not be spending its limited funds on national stormwater programs until it had resolved the basic question of access to safe water. It could, however, be promoting local solutions to stormwater and use of affordable and appropriate technologies.

While the importance of water governance is not apparent to everyone, it is an essential input to sustainable water management systems. When water governance is effective, we can expect harmony in river basins, shared decisions about use water resources, and common action to respond to emergencies. When it is ineffective or even absent, we expect water conflicts, pollution, and more frequent water emergencies that lack coherent response mechanisms.

Ideally, water management and governance can promote ideals such as stewardship and IWRM, but experience shows that strong incentives and controls are required, even when resources for effective water management are available. The role of water governance is to fill the gaps and empower all aspects in our continuing quest for sustainable water management.

GOOD GOVERNANCE AND TOTAL WATER MANAGEMENT

To close a discussion of a complex topic like water governance it would be nice to sum it up succinctly. However, there are too many facets to the topic to do that in words other than glib generalizations. I'd like to turn to the work of a group that defined Total Water Management and simply outline how good governance is required to deliver the results they sought.

The group of 30 was drawn from the U.S. waterworks industry and is similar to the many, many work groups that grapple with water issues around the world. Although there were no environmentalists, their view was heard loud and clear. Although there were no representatives of the poor, the group was sensitive to their needs. I was the only academic, and I was impressed with the knowledge and commitment of this group of professionals.

The definition they developed and how its elements require good governance is outlined below (Grigg, 2008). The elements of the definition are divided into discrete parts so they can be linked with governance requirements. As you will see, the definition is more about what it should do than about how to do it.

"Total Water Management is:

the exercise of stewardship of water resources for the greatest good of society and the environment.	Emphasis on everyone's responsibility toward shared, social goals
A basic principle of Total Water Management is that the supply is renewable, but limited, and should be managed on a sustainable use basis.	Emphasis on the need for careful management of the limited resource
Taking into consideration local and regional variations, Total Water Management: encourages planning and management on a natural water systems basis through a dynamic process that adapts to changing conditions;	Emphasis on use of natural water systems as a way to organize an adaptive process of management

balances competing uses of water through efficient allocation	Emphasizes the allocation process as a way to balance water uses
that addresses social values, cost effectiveness, and environmental benefits and costs;	Emphasizes the need for the TBL approach to decision making
requires the participation of all units of government and stakeholders in decision-making through a process of coordination and conflict resolution;	Emphasizes broad involvement of stakeholders
promotes water conservation, reuse, source protection, and supply development to enhance water quality and quantity;	Emphasizes the tools of IWRM that must be used to facilitate positive outcomes
and fosters public health, safety, and community good will."	Re-emphasizes the importance of TBL goals

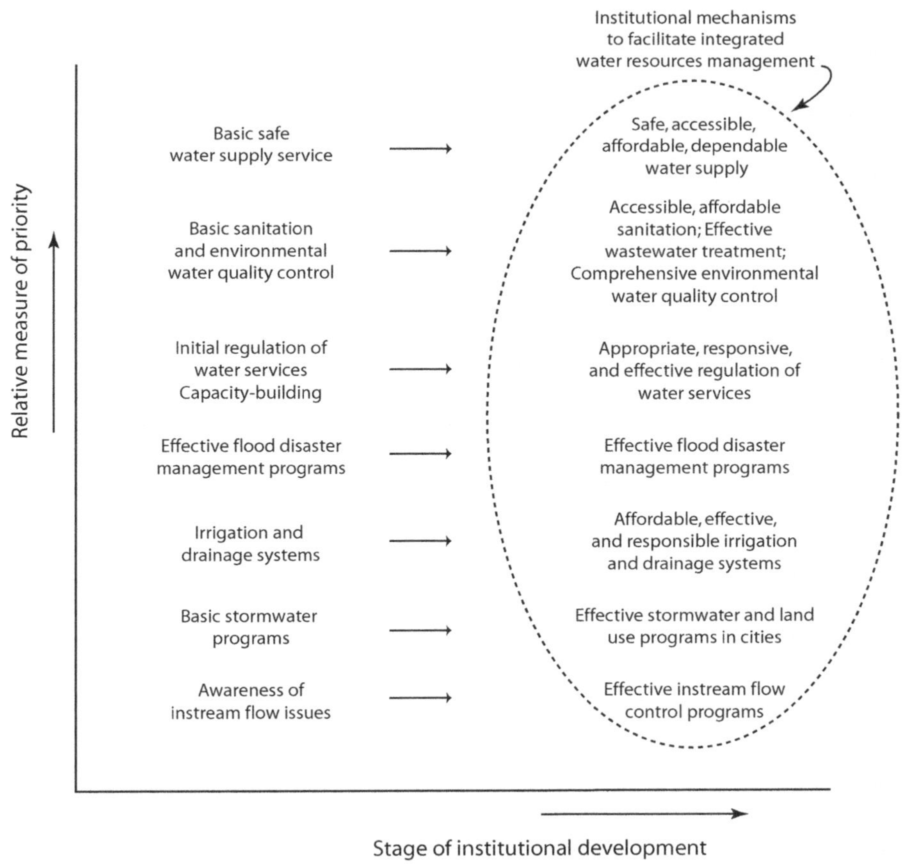

Figure 18.2 A future for integrated water resources management?

As you can see, the policy, capacity-building and control features of water governance must be used to bring about the elements of Total Water Management, which are able to deliver balanced use of water for the greatest good of society and the environment.

FINAL WORD

So, after a long discussion of water governance, what does it add up to? It points to our values, attitudes and actions in recognizing that unless society's institutions function effectively, we cannot meet human and environmental water needs sustainably. Political and social institutions must function well to provide for the balance and order required to manage water systems and organizations. There must be an ethic for stewardship to care for the environment so the Tragedy of the Commons can be prevented. We must cooperate and make tradeoffs that will be good for all. Finally, people must care for each other and share water resources.

REFERENCES

Global Water Partnership (September 2007). *Toolbox: Institutional Setting of Water Institutions in Development, Enforcement and Implementation of Future Water Management Plans.* Bratislava.

Grigg N. S. (2008). *Total Water Management: Leadership Practices for a Sustainable Future.* American Water Works Association Press, Denver.

Seppälä O. T. (2004). *Visionary Management in Water Services: Reform and Development of Institutional Frameworks.* Ph.D. Thesis. Publication 457. Tampere University of Technology.

Index

IWA Publishing's authorised EU representative for General Product Safety Regulations is Diane D'Arras, 15 rue Duret, 75116 Paris, France, e-mail: safety@iwap.co.uk.

Printed and bound by CPI Group (UK) Ltd, Croydon, CR0 4YY
12/05/2026
02108863-0008